Sport To   oment

ASPECTS OF TOURISM

Series Editors: **Chris Cooper** *(Leeds Beckett University, UK)*, **C. Michael Hall** *(University of Canterbury, New Zealand)* and **Dallen J. Timothy** *(Arizona State University, USA)*

Aspects of Tourism is an innovative, multifaceted series, which comprises authoritative reference handbooks on global tourism regions, research volumes, texts and monographs. It is designed to provide readers with the latest thinking on tourism worldwide and in so doing will push back the frontiers of tourism knowledge. The series also introduces a new generation of international tourism authors writing on leading edge topics.

The volumes are authoritative, readable and user-friendly, providing accessible sources for further research. Books in the series are commissioned to probe the relationship between tourism and cognate subject areas such as strategy, development, retailing, sport and environmental studies. The publisher and series editors welcome proposals from writers with projects on the above topics.

All books in this series are externally peer-reviewed.

Full details of all the books in this series and of all our other publications can be found on http://www.channelviewpublications.com, or by writing to Channel View Publications, St Nicholas House, 31-34 High Street, Bristol BS1 2AW, UK.

ASPECTS OF TOURISM: 84

Sport Tourism Development

3rd Edition

James Higham and Tom Hinch

CHANNEL VIEW PUBLICATIONS
Bristol • Blue Ridge Summit

DOI https://doi.org/10.21832/HIGHAM6553

Library of Congress Cataloging in Publication Data
A catalog record for this book is available from the Library of Congress.
Names: Higham, James E.S., author. | Hinch, Thomas, author.
Title: Sport Tourism Development/James Higham and Tom Hinch.
Other titles: Aspects of Tourism.
Description: Third edition. | Blue Ridge Summit, Pennsylvania:
   CHANNEL VIEW PUBLICATIONS, [2018] | Series: Aspects of Tourism: 84 |
   Previous edition: 2011. | Includes bibliographical references and index.
Identifiers: LCCN 2017054575| ISBN 9781845416546 (pbk : alk. paper) |
   ISBN 9781845416553 (hbk : alk. paper) | ISBN 9781845416560 (pdf) |
   ISBN 9781845416577 (epub) | ISBN 9781845416584 (kindle)
Subjects: LCSH: Sports and tourism.
Classification: LCC G155.A1 H56 2018 | DDC 338.4/791--dc23 LC record available at
https://lccn.loc.gov/2017054575

British Library Cataloguing in Publication Data
A catalogue entry for this book is available from the British Library.

ISBN-13: 978-1-84541-655-3 (hbk)
ISBN-13: 978-1-84541-654-6 (pbk)

Channel View Publications
UK: St Nicholas House, 31-34 High Street, Bristol, BS1 2AW, UK.
USA: NBN, Blue Ridge Summit, PA, USA.

Website: www.channelviewpublications.com
Twitter: Channel_View
Facebook: https://www.facebook.com/channelviewpublications
Blog: www.channelviewpublications.wordpress.com

The policy of Multilingual Matters/Channel View Publications is to use papers that
are natural, renewable and recyclable products, made from wood grown in sustainable forests. In the manufacturing process of our books, and to further support our
policy, preference is given to printers that have FSC and PEFC Chain of Custody
certification. The FSC and/or PEFC logos will appear on those books where full
certification has been granted to the printer concerned.

Typeset by Deanta Global Publishing Services Limited.
Printed and bound in the UK by Short Run Press Ltd.
Printed and bound in the US by Edwards Brothers Malloy, Inc.

Contents

Figures and Tables

Case Study and Focus Point Contributors

Mike Boyes. School of Physical Education, Sport and Exercise Sciences, University of Otago, Dunedin, New Zealand. Email: mike.boyes@otago.ac.nz

Inge Derom. Department of Sport Management and Policy, Vrije Universiteit, Brussels, Belgium. Email: inge.derom@vub.be

Adam Doering. Centre for Tourism Research, Faculty of Tourism, Wakayama University, Wakayama, Japan. Email: adoering@center.wakayama-u.ac.jp

Debbie Hopkins. Department of Geography and the Environment, University of Oxford, Oxford, United Kingdom. Email: debbie.hopkins@ouce.ox.ac.uk

Eiji Ito. Faculty of Tourism, Wakayama University, Wakayama, Japan. Email: eijito@center.wakayama-u.ac.jp

Daniel Evans. Department of Geography, York University, Toronto, Canada. Email: devans05@yorku.ca

Scarlett Hagen. School of Physical Education, Sport and Exercise Sciences, University of Otago, Dunedin, New Zealand. Email: scarlett.hagen@otago.ac.nz

Millicent Kennelly. Department of Tourism, Sport and Hotel Management, Griffith University, Nathan, Australia. Email: m.kennelly@griffith.edu.au

Brendon Knott. Department of Sport Management, Cape Peninsula University of Technology, Cape Town, South Africa. Email: brendonknott@gmail.com

Cory Kulczycki. Faculty of Kinesiology and Health Studies, University of Regina, Saskatchewan, Canada. Email: Cory.Kulczycki@uregina.ca

Matthew Lamont. School of Business and Tourism, Southern Cross University, Lismore, Australia. Email: matthew.lamont@scu.edu.au

Brent Moyle, Griffith Institute for Tourism, Griffith University, Nathan, Australia. Email: b.moyle@griffith.edu.au

Glen Norcliff. Department of Geography, York University, Toronto, Canada. Email: gnorcliff@york.ca

Gregory Ramshaw. Parks, Recreation & Tourism Management, Clemson University, Clemson, South Carolina. Email: gramsha@clemson.edu

Arianne C. Reis. School of Science and Health, Western Sydney University, Penrith, Australia and Adjunct Research Fellow, School of Business and Tourism, Southern Cross University, Lismore, Australia. Email: A.Reis@westernsydney.edu.au

Michelle Rutty. Department of Community Sustainability, Michigan State University, Lansing, Michigan. Email: mrutty@anr.msu.edu

Richard Shipway. Department of Sport and Physical Activity, Bournemouth University, Poole, United Kingdom. Email: RShipway@bournemouth.ac.uk

Robert Steiger. Department of Public Finance, University of Innsbruck, Innsbruck, Austria. Email: Robert.Steiger@uibk.ac.at

Acknowledgements

This third edition, like the two that preceded it, has only come to fruition due to the encouragement, contributions and support of many friends and colleagues. We would like to acknowledge the support of Sarah Williams, Elinor Robertson and Flo McClelland (Channel View Publications) and the Aspects of Tourism series editors, Chris Cooper (Leeds Beckett University), Michael Hall (University of Canterbury) and Dallen Timothy (Arizona State University). Work on the planning and writing of this third edition took place during periods of study leave in February 2017 (Dunedin, New Zealand) and June 2017 (Edmonton, Canada) for which we acknowledge the support of the University of Otago and the University of Alberta. James also acknowledges the support of the University of Stavanger (Norway) Visiting Professorship and the University of Queensland (Australia) Jim Whyte Fellowship, and Tom his position at Wakayama University (Japan) as a visiting Distinguished University Professor.

We are grateful for the case study contributions of Richard Shipway (University of Bournemouth), Eiji Ito (Wakayama University), Brendon Knott (Cape Peninsula University of Technology), Arianne Reis (University of Western Sydney), Daniel Evans (York University), Debbie Hopkins (University of Oxford), Brent Moyle (University of the Sunshine Coast), Robert Steiger (University of Innsbruck) and Adam Doering (Wakayama University) and the focus point contributions of Cory Kulczycki (University of Regina), Greg Ramshaw (Clemson University), Inge Derom (Vrije Universiteit Brussel), Millicent Kennelly (Griffith University) Matt Lamont (Southern Cross University), Glen Norcliff, (York University), Michelle Rutty (Michigan State University) and Scarlett Hagen and Mike Boyes (University of Otago). Their critical insights have stimulated our work in the field of sport tourism, and contributed theoretical and empirical insights that illustrate and advance the discussions that we present in this edition. We are also indebted to Sabine Parry (University of Otago), Aisulu Abdykadyrova (University of Alberta) and all of our colleagues at the Department of Tourism, University of Otago and the Faculty of Kinesiology, Sport, and Recreation, University of Alberta.

Finally, the support of our immediate families, Linda, Alexandra, Kate and George, and Lorraine, Lindsay and Gillian, has been critical to the completion of this third edition.

James Higham   Tom Hinch
Dunedin, New Zealand   Edmonton, Canada

This book is dedicated to:
Hawea and Lac Lu

Part 1
Introduction

1 Sport Tourism in Times of Change

In terms of popular participation, and in some aspects of practice,
(sport and tourism) are inextricably linked... and there are sound
reasons for those links to strengthen.
Glyptis, 1989: 165

Introduction

In June 336 BC, Philip of Macedon (382–336 BC) made preparations for the wedding of his daughter Cleopatra to King Alexander of Epirus. He intended, according to Green (1992), to make the wedding a lavish and ostentatious display of propaganda, in order to impress the Greeks into reimagining Philip as a 'civilised and generous statesman', rather than a military tyrant and autocrat. Philip gathered his Macedonian barons and invited all manner of distinguish Greek visitors. No expense was spared in hosting banquets and entertaining his visitors with musical performances and extravagant sacrifices to the Gods (Green, 1992). Interestingly, in the context of this book, public games and displays of competitive sporting excellence were other important elements of Philip's strategy, although the celebrations were cut short by his violent assassination.

Two years later, in 334 BC, Philip's son and successor, Alexander the Great (365–323 BC) hosted the 'so-called "Olympian Games", a nine day festival in honour of Zeus and the Muses held at Aegae or Dium' (Green, 1992: 163). His intentions were to dispel rumours of Macedonian financial ruin with splendid banquets hosted in enormous marquees set against the backdrop of Mount Olympus, and to impress and win the favour of his senior officers and Greek city state ambassadors on the eve of his great Persian campaign. But, above all, Alexander sought through hosting the 'Olympian Games' to impress upon all onlookers his (self)image as a golden demigod of Hellenism, beyond mortal scale. Clearly, hosting sport competitions, as well as royalty, politicians, ambassadors and dignitaries, all seeking some form of association with sports events, is historically longstanding (Green, 1992; Keller, 2001).

Sports, particularly large-scale sporting events dating back to the ancient Olympian Games, have long influenced travel (Keller, 2001). However, few definitions of sport adequately express its diverse and dynamic nature and its changing functions in different societies over time. While sports

3

are as old as civilisation (Coakley, 2017), and the defining qualities of sports are well established, few definitions of sport capture the changing place of sports in societies. As Andrews (2006: 1) observes, 'although physically-based competitive activities are a feature of virtually all human civilisations, the popular myth of sport as a fixed and immutable category is little more than a pervasive, if compelling, fiction'. Instead, Andrews (2006) recommends interpretive approaches to understanding sports, whereby the study of sport and sport experiences are firmly anchored in an understanding of socio-historical context. Sports, then, are a reflection of their historical and social circumstances. This point is rooted in Bale's (1989) simple but useful notion that sport is defined by what features on a day-to-day basis in the sports section of local newspapers. The content of any daily newspaper will be a reflection of its historical and social context. If nothing else, scrutiny of the sports pages of local newspapers confirms the vast diversity of sports relative to their situation in place and time (Higham & Hinch, 2009), although the function of newspapers as a local voice has been diluted by the growth of online newsfeeds, blogs and social media that are far less spatially defined.

Alexander's use of the 'Olympian Games' as a display of personal godliness has, since 1896, thrown up many political and commercial parallels in the modern Olympic Games. It is also clear that the scale, complexity and potential of sport tourism, as well as the expanding mutual interests of the sport and tourism industries that have developed as a consequence, demand critical academic attention (UNWTO, 2017). This book is about sport tourism and its manifestations in space and time. It articulates the defining qualities of sport that explain its unique contribution to tourism. It then applies tourism development concepts and themes to the study of sport tourism. Three key questions structure our discussions of sport tourism development in this book: 'What makes sport unique as a tourist attraction or activity?', 'How is sport tourism manifest in space?' and 'How do these manifestations change over time?'

The chapters that comprise this book are organised into five parts. This chapter (Part 1: Introduction) introduces the purpose and structure of the book. It describes the development and growth of sport tourism, and then raises questions that are intended to demonstrate the relevance and challenge the assumptions that the reader may have on this subject. Part 2: Foundations of Sport Tourism Development comprises Chapters 2–4. This section is intended to provide the reader with fundamentals in the study of sport tourism, sport tourism markets and development processes and issues relating to sport tourism. Much progress has been made in the study of sport and tourism in recent years (e.g. Fyall & Jago, 2009; Gammon, 2015; Gammon *et al.*, 2013; Gibson, 2005; Hallmann *et al.*, 2015; Higham & Hinch, 2009; Lamont, 2014; Preuss, 2015; Taks, 2013; Taks *et al.*, 2015; Weed, 2007; Weed & Bull, 2012; Weed *et al.*, 2014). Chapters 2–4 provide an opportunity to review current insights into sport tourism as a basis for the discussions that follow.

Part 3: Sport Tourism Development and Space (Chapters 5–7) focuses on the spatial elements of sport tourism development. These chapters examine sport tourism development in relation to space, place and the environment. Each of these topics represents important geographical aspects of sport and tourism development. Part 4: Sport Tourism Development and Time (Chapters 8–10) examines sport tourism development in relation to time. The short-term, medium-term and long-term time horizons provide a temporal framework that frames our considerations of the immediate sport tourism experience; sport tourism and seasonality; and the dynamic interrelationship between sport and tourism within long-term evolutionary frameworks. Part 5: Conclusions (Chapter 11) concludes by reviewing the preceding discussions to serve as a platform from which to consider the future of sport and tourism, and sport tourism research. This structure provides a framework that raises questions relating to sport tourism development in space and time and to address these questions through the application of relevant theory.

Sport Tourism in Times of Change

The Football World Cup, hosted in recent years by Germany (2006), South Africa (2010) and Brazil (2014), is one of the world's truly mega sports events. It is a month-long showcase of football skill, and a stage upon which collective identities are forged and nationalisms are expressed (Giulianotti, 1995a, 1995b, 1996). Indeed, it is a stage that is shared by individual players and national teams, as well as politicians, civic leaders, multinational corporate actors, media corporations, spectators and tourists (Cornelissen, 2010). Events such as the Football World Cup and the Olympic Games represent the apex of elite competitive sport. However, they exist upon a superstructure of ever-expanding and diversifying participation in sport and recreational activity, and as one part of an ever-evolving global tourism industry. Once, sport-related international travel was the domain of elite athletes, representing their countries in international competition. With expanded personal mobilities in developed societies has come sport-related travel that extends across all spatial scales, levels of competition and competitive–participatory, serious–casual and active–passive dimensions of engagement (Higham & Hinch, 2009). Thus, the Football World Cup and mega sports events more generally, must be recognised as but one of the manifold diverse forms of sport-related tourism.

Democratisation, the process of opening access to previously restricted opportunities, applies to the development of sport and tourism in recent decades (Standeven & De Knop, 1999). Participation in some sports remains defined by factors such as social class; 'Irrespective of culture or historical period, people use sport to distinguish themselves and to reflect their status and prestige' (Booth & Loy, 1999: 1). The existence of post-class egalitarian consumer cultures in sport is balanced by the fact that similar status groups generally share lifestyle and consumption patterns (Booth & Loy, 1999). The links that exist between socio-demographic status, lifestyle and consumption patterns in sport and tourism heighten the value and utility of defining sport tourism markets in practice.

That said, the forces of globalisation (Bernstein, 2000; Milne & Ateljevic, 2004; Thibault, 2009) and democratisation (Standeven & De Knop, 1999) have had significant implications for the consumption of sport and development processes in sport tourism (Chapter 4). The modern development of sport tourism, then, stands at the cross section of a wide range of contemporary trends in sports participation and tourism development (Table 1.1).

These processes have been driven by neoliberal economic and global political forces (Collins, 1991; Cooper *et al.*, 1993; Gibson, 1998a; Nauright, 1996), as well as changing social attitudes and values (Jackson *et al.*, 2001; Redmond, 1991). They have also been facilitated by technological advances, such as satellite television broadcasting and internet live streaming that have

Table 1.1 Contemporary trends in sports participation and tourism development

Sports participation

(1) The expanding demographic profile of participants in sports (Glyptis, 1989).
(2) Heightened interests in health and fitness in Western societies since the 1970s (Collins, 1991).
(3) Increasing demand for active engagement in recreational pursuits while on holiday since the 1980s (Priestley, 1995; Standeven & De Knop, 1999).
(4) Professionalisation of power and performance sports (Coakley, 2017).
(5) Rapid growth of participation and pleasure sports (Coakley, 2017).
(6) Expanding participation in lifestyle sports with interests in personal and collective identity formation (Gilchrist & Wheaton, 2011; Wheaton, 2004).
(7) Widening scope of performance sports participation beyond elite professional sportspeople, to now include semi-professional athletes and those who engage in amateur sports as a form of serious leisure (Kennelly *et al.*, 2013; Lamont *et al.*, 2014).
(8) Expanding spatial mobilities associated with all forms of engagement in sports (Higham & Hinch, 2009).

Sports and tourism development

(1) Recognition of the association between specific sports and unique tourism destinations (Hinch & Higham, 2004).
(2) Critical insights into the travel flows associated with sport events (Gratton *et al.*, 2006; Preuss, 2005; Weed, 2007).
(3) The role of sports in urban regeneration (Gratton *et al.*, 2005) and destination development (Mason & Duquette, 2008).
(4) The potential contributions of sport to destination image, intention to visit (Chalip *et al.*, 2003; Kaplanidou & Vogt, 2007) and choice of destination (Humphreys, 2011).
(5) The potential for sports to offer destination marketing synergies (Harrison-Hill & Chalip, 2005) through branding (Chalip & Costa, 2005), leveraging (O'Brien & Chalip, 2007) and bundling (Chalip & McGuirty, 2004).

influenced the 'sportification of society' (Halberstam, 1999; Standeven & De Knop, 1999), and the forces of globalisation, which have had fundamental implications for personal and collective identity (re)formation (Higham & Hinch, 2009). The political events of 2016, including the Brexit vote (23 June 2016) and the US presidential elections (8 November 2016), which signalled new directions in economic protectionism and the strengthening of international border processes for visitors and migrants, will have significant implications for sport-related tourism and mobilities over the coming years, particularly sport labour migration. The global sports arena may become less global (Bale & Maguire, 2013), as some demonstrate resistance to the forces of globalisation and show signs of becoming less global in outlook.

Nevertheless, it remains the case that 'the geographical extent and volume of sports related travel has grown exponentially' (Faulkner et al., 1998: 3). Glyptis (1989) was one of the earlier scholars to provide an indication of these trends. She notes in a study of western European countries that all had experienced strong growth in interest in recreational sport during the 1980s. Furthermore, participation was increasing in all social strata, most sports were receiving participants from an expanding social spectrum and all had recorded significant increases in youth holidays, short breaks and multiple annual holidays. The two and a half decades that have followed have served to strengthen these trends (Hall, 1992a, 1992b; International Olympic Committee & World Tourism Organisation, 2001; UNWTO, 2017). Much has continued to change in the study of sport and tourism. Intensifying commercial demands in elite sport, the development of hybrid and innovation of new sports and the growth of lifestyle sports (Gilchrist & Wheaton, 2016; Wheaton, 2004) among other things, have changed the relationship between sport and the environment, sport and place, sports fans and their teams, and sports participants and their constructions of identity (Higham & Hinch, 2009).

The Foundations of Sport Tourism Development

The growth of sport tourism justifies critical consideration of relevant development issues. This task requires that sport tourism is defined and conceptualised in ways that highlight rather than obscure the diversity of interests in sport and tourism. A number of definitions exist in this field, which afford the opportunity to study sport tourism from a variety of perspectives. For the purposes of this book, sport tourism is conceptualised by considering sport as a tourist attraction, and by highlighting the defining qualities of sport that collectively constitute a unique contribution to tourism (Chapter 2: The Study of Sport Tourism). This approach is intended to give prominence to the diverse, dynamic and complex nature of sport and tourism, thereby helping to clarify the parameters and scope of this book.

The diversity of sport tourism markets is explored in Chapter 3: Sport Tourism Markets, which highlights the rich diversity of motivations, and therefore the varied approaches to market segmentation that exist in sport tourism. Bale (1989: 9) notes that 'work-play, freedom-constraint, competition-recreation, and process-product are only some of the continua on which sports can be located'. Thus, the experiences of sport tourists are likely to vary considerably with the motivations that travellers hold towards their chosen sports. The motivations associated with sport tourism niche markets raise intriguing questions for sports event organisers and promoters, sports associations, managers of sporting venues, destination managers and tourism marketers (Higham & Hinch, 2009). To what extent, for example, are highly competitive professional athletes interested in tourist experiences at a sport destination, and how may the potential of this market be fully achieved? The promotional opportunities that derive from the association of high-profile athletes with specific tourist destinations are part of this potential (Chalip, 2004a). How does tourism based on professional sport differ from the tourism development opportunities associated with recreational sport? Tourist experiences relating to sport vary within and between niche market segments, which raises questions as to how these markets can be better understood by sport and tourism managers who seek to meet changing sport and travel preferences.

The logical extension of this market analysis is the consideration of development processes, sustainability and planning interventions. Development issues are of particular interest to sport and tourism practitioners. This is made most evident by the publication of the United Nations World Tourism Organisation (UNWTO, 2017) Global Report on Sport Tourism (Focus Point 1.1). This report was developed with the goal of providing timely and practical insights into such matters as infrastructure and facilities planning, collaboration between sport and tourism agencies, public–private partnerships (PPP) in sport tourism and new trends (and emerging markets) in sport and tourism. Other critical development issues in sport tourism relate to commodification/authenticity and global–local processes. Hitherto, little consideration has been given to the modification of sports competitions (e.g. through rule changes, length and timing of the competition season and the televising/streaming of live sport), or the development of new or hybrid sports, specifically in terms of the potential development opportunities for tourism destinations (Higham & Hinch, 2002a, 2002b). The implications for tourism destinations may include the emergence of new visitor markets, altered seasonal travel flows, elevated or altered perceptions of place, close alignment with the interests of local communities and new elements of destination imagery associated with sports. The relevance of these processes and issues are introduced to the reader, and explored in Chapter 4: Development Processes and Issues. These preliminary chapters provide a foundation upon which the subsequent chapters (Chapters 5–10) build.

Focus Point 1.1: UNWTO Global Report on Sport Tourism

In 2017, the United Nations World Tourism Organisation (UNWTO) announced that it was preparing the UNWTO Global Report on Sport Tourism in a project intended to provide systemic insights into the multifaceted and dynamic links between the sport and tourism sectors. Working in collaboration with the Ostelea School of Tourism and Hospitality (Spain), the report set out to overview the manifold possibilities arising from sport, particularly in terms of economic and socio-cultural development in tourism. Furthermore, the report was intended to critically examine the 'challenges and opportunities inherent in Sport Tourism and identify the key elements necessary for the successful design and implementation of development strategies in Sport Tourism, both from a public and private perspectives' (UNWTO, 2017 online). Among other things, the UNWTO Global Report on Sport Tourism was undertaken to address:

(1) Local resident participation in the design of sport tourism programmes.
(2) Infrastructure and facilities planning and development for sport tourism.
(3) Collaboration between sport and tourism agencies.
(4) Successful PPP in sport tourism.
(5) Social and economic development through sport activities.
(6) Sport tourism activities as a driver of local development.
(7) Sport tourism and overcoming seasonality challenges at destinations.
(8) Sport tourism and sustainability: overcoming environmental degradation.
(9) Small-scale sport tourism activities and events as mechanisms of local economic development and local identity building.
(10) New trends and emerging markets in sport tourism (e.g. surf, drones, yoga, extreme sports, invention of new styles of sports).
(11) Diversification of products and destinations through sport tourism.
(12) Destination branding and sport tourism markets.
(13) Sport tourism and experiences of culture at destinations.
(14) Sport tourism initiatives relating to health and wellness.
(15) Challenges and opportunities to develop competitive sport tourism destinations.
(16) The role of policymakers and industry in the successful management of sustainable sport tourism facilities.
(17) Loyalty of high-end consumers in sport tourism (e.g. repeat participation in marathons events).

(Continued)

Focus Point 1.1: (Continued)

The UNWTO Global Report on Sport Tourism provides insights into sport tourism phenomena in relation to policy and planning for development, and highlights the expanding intergovernmental interest in sport and tourism in relation to economic and socio-cultural development at the global scale.

Source: UNWTO (2017) http://affiliatemembers.unwto.org/publications (accessed 12 June 2017).

Sport Tourism Development and Space

Chapters 5–8 consider how sport tourism is manifest in space and how these manifestations may be influenced. Chapter 5: Space: Location and Travel Flows explores the interrelationships linking sport tourism-generating areas and destinations, and the travel patterns associated with sport tourism markets. The basic concepts and themes for this chapter have their roots in economic geography. These concepts are drawn from the study of sports geography and the spatial analysis of sports (Bale, 1989, 1993), which inform, for example, the allocation of professional sport team franchises within a sports league or decisions on where to build, develop or enhance sport resources and facilities. The ways in which sports may influence the spatial travel patterns and itineraries of visitors travelling to and within a destination, whether sport functions as a primary, secondary or tertiary attraction, are also discussed. The rapidly expanding world of e-sports and virtual reality in sport experiences raises interesting questions regarding the relationship between sport and space.

The sports played in a region influence the meanings that are attached to space. Concepts of place, culture and place promotion are explored in Chapter 6: Place, Sport and Culture. In many ways, sport infuses tourism spaces with one of the most authentic types of attractions. The link between culture and sport takes many forms. In Chapter 6, we explore the notions of 'sport and culture', 'sport as culture' and 'sport subcultures'. All of these variations contribute to the meaning attached to sport tourism places (Focus Point 1.2). Strategies that incorporate these cultural variations can be used to promote place to a variety of markets. There are, however, significant challenges associated with the commodification of culture in sport (Jackson *et al.*, 2001).

Focus Point 1.2: Sport, Heritage and Culture

Heritage and culture contribute to making tourism destinations unique. Sport and sports venues may represent a unique expression of the heritage of place. Sports venues in the UK, such as Lord's (cricket),

Wimbledon (lawn tennis), Twickenham (rugby union), St Andrews (golf), Wembley (football) and Royal Ascot (horse racing) are widely recognised as the spiritual homes of their sports. Some of sport's most prestigious tournaments are contested at these venues. With time, each has developed its own aura of tradition. Each represents a significant expression of the heritage and culture of Britain (British Tourist Authority, 2000) and many have developed museums, halls of fame and venue tours to foster visitor experiences and to cement this spiritual status. Other sports represent distinctive cultural elements of a regional or national destination. The examples are universal and include football (Brazil), ice hockey (Canada), baseball (USA), Australian Rules football (Australia), Rugby Sevens (Fiji), sumo wrestling (Japan) and Thai boxing (Thailand).

These sports allow ready access to the culture of a destination. In Brazil, football legend is built upon 'ginga', a term used by Pele (three times FIFA World Cup winner and World Footballer of the 20th century) from the Portuguese word for 'sway' that came to describe a style of play based on flowing creativity and the rare individual brilliance that arises in some Brazilian players. In Fiji, Rugby Sevens is an expression of the flair and style of Melanesian and Polynesian rugby, combining individual speed and strength with a willingness to play a fast pace, high risk passing game. The brilliance of Fijian Rugby Sevens culminated in Rio Olympic gold in 2016; a remarkable feat for a small and poorly resourced Pacific Island nation. In reference to Māori *haka* (challenge), Laidlaw (2010) notes that since the 1980s the All Blacks performance of *haka* before rugby internationals has been performed with 'genuine cultural meaning'. This has included the creation of a new All Blacks *haka*, *Kapa o Pango* (translated to 'the group in black'), to sit alongside the traditional *haka* (*Ka Mate*). 'It's evolution mirrors perfectly the cultural renaissance of the Māori in New Zealand' (Laidlaw, 2010: 186). The performance of *haka* by sports teams, including the All Blacks, contributes to both national identity and tourist experiences, representing one of the more prominent, unique and important parts of the cultural experiences of visitors to *Aoetaroa* New Zealand (see also Jackson *et al.*, 2001).

Key Reading

Jackson, S.J., Batty, R. and Scherer, J. (2001) Transnational sport marketing at the global/local nexus: The Adidasification of the New Zealand All Blacks. *International Journal of Sports Marketing and Sponsorship* 3 (2), 55–71.

Laidlaw, C. (2010) *Somebody Stole My Game*. New York: Hachette.

The environment is the subject of the third spatial dimension (Chapter 7: Environment: Landscapes, Resources and Impacts). This chapter considers the resource base for sport and tourism facilities and infrastructure. Quite different issues are associated with natural resources and built facilities in sport tourism. Outdoor sports such as downhill skiing and snowboarding, for example, tend to be dependent on specific types of landscape, with the potential for environmental impacts. Other types of sport are more transportable and feature standard facilities that can be built in locations designed to maximise market access. The development of built resources for sport, and the shift towards artificial, enclosed and controlled sports environments, has been a significant trend for the last two decades. The question is, to what extent can nature be removed from sports, and sports from nature, and will there be a backlash? It has been interesting to observe academic exploration of the interplay of sports and nature, for example, in influencing sports participants to better understand human relationships with the natural environment (Krein, 2008).

Sport Tourism Development and Time

Chapters 8–10 consider the manifestations of sport tourism in time. Chapter 8: Sport and the Tourist Experience explores the short-term temporal horizon of sport tourism. The visitor experience is concerned with the timing and duration of visits, engagement in sports and touristic and leisure activities at the destination. Different forms of sport tourism manifest themselves in contrasting tourist experiences. Sport experiences derived from spectator events, competitive engagement in events, active participation in recreational sports and sports heritage and nostalgia give structure to the discussions presented in Chapter 8. This framework lends itself to consideration of the co-creation of sports experiences at a destination (Morgan, 2007), and situates the experience of sport tourism within Weed's (2005) interplay of activity, people and place. Chapter 8 also examines the relevance of the sport and tourism systems that mediate the sport tourist experience at a destination. This approach is intended to provide insights into strategies that may influence and enhance sport tourism visitor experiences.

The medium term or seasonal dimension of sport tourism is the subject of Chapter 9: Seasonality, Sport and Tourism. Few tourism destinations are unaffected by systematic seasonal fluctuations in the tourism phenomenon. Little is known about how climate change is giving rise to less predictable seasonal weather patterns, contributing to regional vulnerabilities in terms of the expanding or contracting availability of seasonal sports resources (Hopkins et al., 2013). Strategies designed to extend shoulder seasons or create all-season destinations are commonplace (Hudson & Cross, 2007). Therefore, given the uncertainties of climate change in relation to sport and tourism (Hopkins, 2014), the manner in which sports moderate, or may be engineered to alter seasonal patterns of visitation (Higham & Hinch, 2002a, 2002b), is now more important than ever. The reverse also applies, whereby tourism may influence patterns of seasonal participation in sport. It is not enough to know that there are seasonal patterns of sport and tourism, it is also important to understand the reasons for those patterns. Leisure constraints theory provides insights into such patterns (Hinch & Jackson, 2000). Consideration can then be given to strategies, such as facility design, pricing and promotions, and event production that may moderate or alter patterns of seasonality in sport and tourism (Focus Point 1.3).

Focus Point 1.3: Sport and the Subjugation of Nature

Increasingly, built sports resources are becoming removed from nature. Enclosed stadiums simultaneously represent the protection of sports from weather-related interruption, and the removal of sports from unique place-specific weather conditions. In 2010, Wimbledon revealed the completion of a roof enclosing the historic centre court, departing from the continuing tradition at the US and French Opens where the Grand Slam tournaments remained outdoors (Melbourne Park, home of the Australian Open has a retractable roof which is often used to protect players from the heat of the sun). While Wimbledon has regularly been subject to rain interruptions, there are elements of nostalgia associated with weather delays, which allow for a pint of lager on the hill, or strawberries and cream while the players take shelter, and the ball boys and girls bring the covers onto Centre Court. Weather interruptions have also played a part in the ebb and flow of historic games, and influenced the outcome of matches in some unforgettable chapters of the tournament's history.

The technology of modern stadiums and arenas allows many sports to be performed without interference or interruption caused by the natural elements. The extent to which sports may become detached from the influences of the weather is, in many cases, so complete that sports managers may moderate the seasonal context of sports. The retractable stadium roof has significant implications for sport and tourism seasonality. Indeed, the first international rugby and cricket fixtures contested indoors took place out of season at Millennium Stadium, Cardiff (Wales) in October 1999 and Colonial Stadium, Melbourne (Australia) in July 2001, respectively. The former features a playing surface that rests on pallets which can be removed and replaced when necessary. Following these vanguard examples, the cases of enclosed stadiums have proliferated worldwide. For example, Sapporo, situated in Hokkaido, Japan's northern-most island, features a futuristic dome, where World Cup matches were played in 2002. The Sapporo Dome is an all-weather covered stadium designed with a

(Continued)

Focus Point 1.3: (Continued)

view to the local climate conditions, particularly heavy snow in winter. This hi-tech facility, which combines indoor and outdoor arenas and an unprecedented hovering football stage, makes it possible to play at any time of year regardless of the weather. The natural grass playing surface can be moved in and out of the stadium, being kept outside the dome to allow the grass to grow when not in use and then moved inside when needed. The entire lower section of the field is rolled into the dome on a cushion of air. As it does so, a rotating seat system moves aside before the pitch turns sideways on its axis and the seating areas automatically slide back into place. The entire manoeuvre takes two hours to complete. The air inside is moderated and controlled by air-conditioning and a natural ventilation system in summer, while spectators are kept comfortable in winter by a heating system applied directly to the seats. The 'hovering stage' and 'moving wall' features allow the natural turf soccer pitch to be replaced with a baseball diamond and outfield, with the open area outside the stadium used to grow the turf when the stadium is not in use.

Source: http://fifaworldcup.yahoo.com/en/da/c/sapporo.html (accessed 28 August 2012).

Chapter 10: Evolutionary Trends in Sport Tourism examines the interrelationship between sport and tourism within an evolutionary or long-term context. Tourism development processes, as conceptualised in the evolution in tourism destinations through a life cycle (Butler, 1980), may be influenced by the powerful dynamics of sport. For example, evolving spatial patterns of sport may have a direct bearing on tourism development. The reverse is also true, as tourism may impact upon the types of sports practiced in destination areas. Surfing serves as a good illustration of this process given its diffusion from its spiritual home of Hawai'i into new regions throughout the world in response to local and non-local (tourist) demand. Heritage sport tourism is a unique form of tourism, which over time may offer tourism development opportunities associated with the search for sporting experiences associated with earlier periods (Gammon & Ramshaw, 2007).

The last decade has confirmed a surge in demand for heritage in sport (Ramshaw & Gammon, 2016). Perhaps this is a response to the growing dominance of commercial interest in sport, reigning over the interests of fans and participants (Laidlaw, 2010). Clearly, sport is a dynamic field of study in terms of its manifestations in space and time and the interplay between sport and tourism. For example, the challenges associated with

SPORT TOURISM

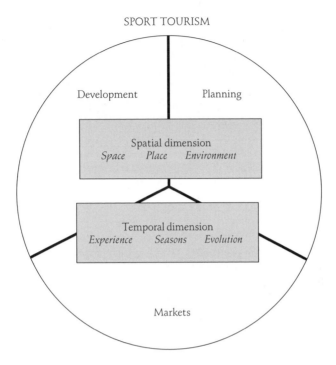

Figure 1.1 Conceptual framework providing the book's structure

commodification – finding a balance between progress and tradition – and contrasting local–global interests in sport need to be systematically explored with direct reference to the study of tourism.

Figure 1.1 presents a framework that provides the structure for the chapters that follow. These spatial (space, place and environment) and temporal (experience, seasons and evolution) themes form the structure of Chapters 5–7 (Part 3) and 8–10 (Part 4), respectively. The foundations of the book, then, lie in the geographical principles of space, place and environment. The application of these themes to sport tourism requires a multidisciplinary approach. This book draws from the fields of sport management; the sociology of sport; consumer behaviour; sports marketing; economic, urban and sports geography; and tourism studies in discussing the manifestations of sport tourism development in space and time. To illustrate points of discussion, Chapters 2–10 include case studies provided by leading and emerging scholars in the study of sport and tourism from a range of disciplines. 'Focus Points' are also integrated into each of these chapters in an attempt to illustrate key points with real-world examples. The overarching goal is to advance theoretical thinking on the subject of sport tourism development generally, and critical thinking on the interplay of local and global forces in sport and tourism development.

Part 2
Foundations of Sport Tourism Development

2 The Study of Sport Tourism

From the stand point of theory, it is necessary to understand what sport tourism shares with, and what distinguishes it from other touristic activities.
Green and Chalip, 1998: 276

Introduction

The phenomenon of sport-related tourism and the implications of these activities for people and places, justify targeted scholarly attention. Theoretical and empirical research offers the opportunity to examine untested assumptions, develop an understanding of complex relationships and processes, and influence outcomes resulting from the interplay of sport and tourism. A focused approach to studying sport tourism will capture synergies leading to new insights into this phenomenon that would not otherwise emerge. Chapter 2 is an articulation of this argument. It examines the conceptual foundations of the study of sport and tourism, arguing in support of directed scholarly attention and consideration of sport within a tourist attraction framework. Richard Shipway's case study of Bournemouth University's well-established and successful sport tourism programme highlights the development and merit of dedicated sport tourism units in tertiary academic degrees (Case Study 2.1).

Conceptual Foundations

The logical starting point to understanding the confluence of sport and tourism is to articulate the essence of each of the parent fields of study. This is not a simple task, as there are multiple perspectives of each realm (Hinch & Higham, 2001). A range of definitions for sport and tourism can be justified in different contexts. In revisiting the concept of sport, Hsu (2005) categorised efforts to define sport as narrow (closed) or broad (open). In the first instance, sport is seen as something that can be demarcated from non-sport activities while the former suggests that this distinction is not always possible. Commentators such as Andrews (2006) align with the latter view, warning that definitions that try to delineate socially constructed phenomena are bound to be futile as social constructions vary across time and space. While agreement on a single universal definition is unlikely and perhaps undesirable, it is helpful for readers to understand the perspectives that have been adopted for this book. The basic parameters that follow are not meant to deny or diminish other perspectives but rather to position our discussions.

Domain of sport

The popular perception of sport as articulated in the *Oxford English Dictionary* (2017) is 'an activity involving physical exertion and skill, especially (particularly in modern use) one regulated by a set rules or customs in which an individual or team competes against another or others'. While reflecting popular sentiment, this definition is consistent with key elements in the scholarly definitions that are found in the field of sport sociology. For example, Woods (2016) also highlights characteristics like competition, physical activity and rules. Such elements are consistent with earlier definitions that defined sport as 'a structured, goal-oriented, competitive, contest-based, ludic physical activity' (McPherson *et al.*, 1989: 15). Increasingly, however, definitions of sport carry a caveat that the meaning of sport varies in terms of time and place.

Basic parameters that emerge from this perspective include rules, competition, play and physical activity. Rules typically relate to space and time. They are observable in a variety of ways including the dimensions of the playing area and the duration and flow of the game or contest. They tend to be more specific in formal variations of a sport, especially as the level of competition increases. Rigid and often complex codes of rules normally govern international competition at an elite level while the rules may be very flexible and general for informal activities. Examples of the latter include the unwritten etiquette of surfing (Usher & Gomez, 2016) or the simple rules agreed to during a spontaneous game of football during a grade school recess.

Sport is also characterised by being goal-oriented, competitive and contest-based. All three characteristics are closely related. Sport is goal oriented in the sense that sporting situations usually involve an objective for achievement in relation to ability, competence, effort, degree of difficulty, required skill set and mastery or performance. In most instances, this goal orientation is extended to some dimension of competition. At one extreme, competition is expressed in terms of winning or losing. Alternatively, competition can be interpreted much less rigidly in terms of competing against standards, inanimate objects, forces of nature or oneself. In the context of sport tourism, the latter interpretation of competition offers a much more inclusive approach that covers recreational sports such as those commonly associated with outdoor pursuits. It is also inclusive of leisure sports (Spracklen, 2013) and contemporary sports such as skateboarding, kite-surfing and others (Wheaton, 2013). Competition is best conceptualised as a continuum that ranges from recreational to elite. Closely associated with competition is the contest-based nature of sport in which outcomes are determined by a combination of physical prowess, game strategy and, to a greater or lesser degree, chance.

The third parameter of sport that arises from the McPherson *et al.* (1989) definition is its 'ludic' or playful nature, a term that is derived from the Latin word *ludus*. This component of the definition states that sport is rooted in, although not exclusive to the concept of play. Those activities that are seen as pure work would not normally be considered sport but the presence of some degree of work, in and of itself, does not rule out an activity as sport. Professional sport therefore fits this definition, as does recreational sport. The presence of play in sport is accompanied by uncertain outcome and sanctioned display. Uncertain outcome helps to maintain suspense throughout a sporting engagement and by doing so, it presents unique advantages in terms of tourism authenticity (Chapter 4). Sanctioned display tends to emphasise the exhibition of athletic skills and as such, it broadens the scope of involvement to spectators as well as participating athletes.

Finally, underlying all of these parameters of sport is its physical and kinaesthetic nature. While different sports require different configurations of fine and gross motor movement, physical movement is the most universally recognised characteristic of sport. Physical prowess consists of physical speed, stamina, strength, accuracy, flexibility, balance and coordination. It forms a major part of embodied sport (Wellard, 2016). Beyond these parameters, sport is characterised by travel. Not only do the competitive hierarchies that lead to high performance sport require travel, but so do many recreational sports like snowboarding (mountains) and surfing (surf breaks). Once an athlete moves beyond the early development stages of his or her sport, he or she is likely to become a recurrent traveller.

Domain of tourism

Typically, tourism definitions include those associated with the popular usage of the term (e.g. Simpson & Weiner, 1989), those used to facilitate statistical measurement (e.g. World Tourism Organisation, 1981) and those used to articulate its conceptual domain (e.g. Netto, 2009). Definitions arising from all of these perspectives tend to share three key dimensions. The most prevalent of these is a spatial dimension (Dietvorst & Ashworth, 1995). To be considered a tourist, individuals must leave and then eventually return to their home. While the travel of an individual does not in itself constitute tourism, it is one of the necessary conditions. A variety of qualifiers have been placed on this dimension including a range of minimum travel distances, but the fundamental concept of travel is universal.

A second common dimension involves the temporal nature of tourism. Tourist trips are characterised by a 'temporary stay away from home of at least one night' (Leiper, 1981: 74). Definitions developed for statistical purposes are more specific in terms of the duration of the trip, with the

United Nations (2008) defining visitors in a destination for less than a year and more than 24 hours or a night. Visitors in a country for less than 24 hours or a night are categorised as excursionists. However, this view can be critiqued. Scholars in the field of mobilities criticise these temporal parameters, arguing that in many societies, they have become too limiting. For example, today's transport technologies allow people to travel extensively within the scope of 24 hours and many people travel for more than one year (Hall, 2004).

A third common dimension of tourism definitions concerns the purpose or the activities engaged in during travel and it is within this dimension that many subfields of tourism research find their genesis (e.g. ecotourism and adventure tourism). Of the three dimensions, this is perhaps the one characterised by the broadest range of views. For example, popular dictionary interpretations of tourists tend to focus on leisure as the primary travel activity (e.g. Simpson & Weiner, 1989), while definitions developed for statistical and academic purposes tend to include business activities (United Nations, 2008). One of the reasons people travel is to engage in sport whether it is to the other side of town or the other side of the world.

Conceptualising sport tourism

The study of sport tourism focuses on the extensive intersection of the sport and tourism domains. As such, sport tourism definitions are challenged by the variety and scope of the overlap. The question of whether sport is the primary or secondary motivation for a trip provides one example of these challenges. To deal with this issue, Gammon and Robinson (2003) make a distinction between sport tourists (sport as the primary motivation) and tourism sports (sport as a secondary and sometimes even an incidental travel activity). These differing levels of involvement in sport are included under the term 'sport tourism' in this book. Most definitions encompass spectators as well as athletes and recreational as well as elite competitors (Gibson, 1998a; Standeven & De Knop, 1999; Weed, 2009). They also tend to include explicit requirements for travel away from the home environment along with at least an implicit temporal dimension that suggests that the trip is temporary and that the traveller will return home within a designated time, although Higham and Hinch (2009) highlighted variations of the typical short-term travel focus such as that characterised by temporary sport migrants. Somewhat surprisingly, the major limitation of many definitions is that they are not clear on what constitutes sport. This is understandable given the differing social constructions of sport throughout the world, but it is also problematic when one is trying to understand the scope of sport tourism development.

Weed and Bull (2003, 2009) consciously break the pattern of conceptualising 'sports tourism' based on the existing parameters of sport or tourism with the rationale that sport tourism is a unique phenomenon. They see it as more than simply the sum of the separate entities of sport and tourism. Instead, they define it as a 'social, economic and cultural phenomenon arising from the unique interaction of activity, people and place' (Weed & Bull, 2003: 258). While it is agreed that sport tourism is more than the sum of its parts (e.g. see the underlying framework used in Higham and Hinch [2009]), the conceptualisation of sport tourism in this book has consciously combined the dominant parameters of sport and tourism. This approach is part of our own 'social construction' of the term reflecting backgrounds in tourism and a mindful attempt to articulate our understanding of sport. Similarly, the term sport tourism rather than sports tourism is adopted not because the unique characteristics of individual sports are not appreciated but rather to reflect an emphasis on the common elements of sport as a social institution along with the characteristics that distinguish it from other types of tourism activity (see Gibson, 2002).

For the purposes of this book, sport tourism is therefore conceptualised as *sport-based travel away from the home environment for a limited time where sport is characterised by unique rule sets, competition related to physical prowess and play* (Hinch & Higham, 2001). Sport is recognised as a significant travel activity whether it is a primary, secondary or even a tertiary feature of the trip. It is seen to be an important factor in many decisions to travel and one that may feature prominently in the travel experience and the assessment of the experience. This perspective enables the adoption of an approach in which sport is seen as a tourist attraction.

Conceptualising Sport as a Tourist Attraction

Green and Chalip (1998: 276) have noted that 'from the stand point of theory, it is necessary to understand what sport shares with, and what distinguishes it from other touristic activities'. It is our contention that sport tourism attractions are unique. We have therefore adopted tourist attraction theory as a useful framework for gaining insight into these unique aspects of sport tourism. While the idea of sport as a tourist attraction is not new (e.g. Rooney, 1988), the theoretical basis for this claim is more recent (Gibson, 2006; Higham & Hinch, 2009; Hinch & Higham, 2004; Weed & Bull, 2009). Leiper's (1990) systems approach to tourist attractions and his classic framework for tourism (Leiper, 1979) provide a useful foundation for Figure 2.1. Leiper's (1990) original tourist attraction system contained three parts: (1) a tourist, (2) a central element and (3) an informative element. The tourist attraction system was said to exist when these three elements were connected. In Figure 2.1, the equivalent of these three elements are: (1) the sport travellers in the generating region; (2) the sport feature in the destination; and (3) the markers in the generating,

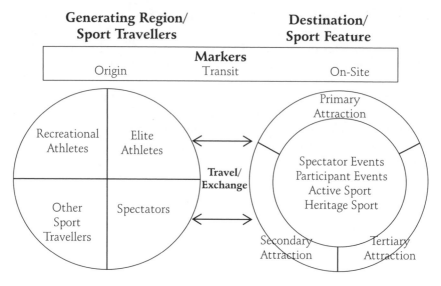

Figure 2.1 The sport tourism attraction system

transit and destination regions that highlight the sporting experiences that are attractive elements to the traveller.

The first, sport travellers, are those who travel away from home to the extent that their behaviour is motivated by or results in sport experiences. Leiper (1990) makes five assertions about the nature of traveller behaviour in general. These hold true for sport travellers with the exception that their motivations and behaviours are focused on sport experiences.

> First, the essence of touristic behaviour involves a search for satisfying leisure [sport] away from home. Second, touristic leisure [sport] means a search for suitable attractions or, to be more precise, a search for personal (*in situ*) experience of attraction systems' nuclear elements. Third, the process depends ultimately on each individual's mental and non-mental attributes such as needs and ability to travel. Fourth, the markers or informative elements have a key role in the links between each tourist and the nuclear elements being sought for personal experience. Fifth, the process is not automatically productive, because [sport] tourists' needs are not always satisfied (these systems may be functional or dysfunctional, to varying degrees). (Leiper, 1990: 371–372)

Sport travellers are characterised by a range of interests which they may choose to pursue one at a time or as is more often the case, in an assortment of combinations. Major segments include: recreational athletes, elite athletes, sport spectators and a variety of related sport travellers such as coaches, officials and media crews interested in high profile sporting events (see Chapter 3).

The second major element of the sport tourism attraction system is the type of sport activity that is of interest in the destination. These manifestations of sport are tied to the location where the sport experience is produced and consumed as well as to interactions with people including local residents and other visitors. More specifically, in the context of sporting attractions, it can typically be classified in terms of site activities related to: (1) spectator-based sport events; (2) participation-based sport events; (3) active sport; and (4) sport heritage activities. This typology of sport tourism activity builds on Gibson's (1998a) typology by distinguishing participation-based sporting events, which can be seen as a hybrid of events and active sport tourism. While spectator events feature a relatively low number of elite competitors and a high number of spectators (e.g. professional sport), participation-based events feature relatively higher proportions of athletes across a spectrum of skill sets and a relatively smaller proportion of spectators (e.g. many triathlons and marathons). The participation-based event category differs from the active sport tourism category in terms of the degree of institutionalisation. Active sport tourism is characterised by direct engagement in a sporting activity such as cycling or tennis inclusive of both recreational and serious involvement. Whereas the sport tourists in the participant-based events (e.g. Gran Fondo group bike ride) depend on event organisers, active sport tourists are relatively less dependent of such organisational structures although they may draw on destination-specific resources such as trail systems and facilities (e.g. independent cycle touring). Finally, heritage sport tourism features are epitomised by attractions such as sport halls of fame but may also include much less tangible resources such as the nostalgia of one's sporting youth. While these categories are useful in the study and practice of sport tourism, they are not exclusive. Major spectator events like the Tour de France may in fact include a combination of spectating and participation events, active and heritage activities depending on the interests and behaviours of individual sport tourists.

Primary attractions are those that have the power to influence a visitor's decision to travel to a destination based solely on that attraction. Secondary attractions are known to a person prior to his or her visit but are not instrumental in and of themselves in the determination of travel itineraries. Tertiary attractions are unknown to the traveller prior to his or her visit but may serve as centres for entertainment, activity or experience once the visitor is at the destination. This hierarchy is evident in sport tourism with many travellers primarily motivated by a particular sporting opportunity (e.g. to watch an Olympic Games), others whose travel decisions depend on a combination of sporting and non-sporting attractions (e.g. a combination of Olympic Games spectating, a golf holiday and some general sightseeing) and still others whose original travel decision may not have been driven by sporting opportunities in the destination but who spend at least part of their visit engaged in destination experiences that are based on sport (e.g. a day of cycle touring during a holiday dominated by visits to historic sites). The sport tourism attraction system, therefore, recognises that sport may function as a tourist attraction in a variety of ways for a variety of people. Appreciating the place of sport within a destination's nuclear mix and hierarchy of attractions, as it relates to different tourist market segments, has significant management implications (e.g. attendance, participation, travel flows, visitor behaviour, timing of visit).

The third element of the sport tourism attraction system consists of markers, which are items of information about any phenomenon that is a potential nucleus element in a tourist attraction (Leiper, 1990). Markers may be detached or removed from the nucleus, contiguous or on-site. In each case, the markers may be positioned consciously or unconsciously to function as part of the attraction system. Examples of conscious attraction markers featuring sport are common. Typically, they take the form of advertisements showing visitors involved in destination-specific sport activities and events. These include advertisements for spectator events such as the Olympics by the host city Destination Marketing Organisation (DMO) and by event sponsors (e.g. Visa, Coca-Cola, McDonald's at the Rio Olympics) who are making substantial investments to link their product to the event brand. Sport images are also very common in other travel advertisements such as resort advertisements that feature scuba divers or kite surfers even though many of the potential tourists will spend the majority of their time sun tanning on the beach. Unconscious detached markers are even more pervasive. At the forefront of these are televised broadcasts of elite sport competitions and advertisements featuring sports in recognisable destinations (Chapter 6). The receivers of this information have the location marked for them as a sport tourist attraction, which may influence future travel decisions.

The conceptualisation of sport tourism used in this book highlights the things that make sport such a popular attraction. First, each sport has its own set of rules that provide characteristic spatial and temporal structures such as the dimensions of a playing surface or the duration of a match (Bale, 1989). Second, competition relating to physical prowess encompasses the goal orientation, competition and contest-based aspects of sport (McPherson et al., 1989). Third, sport is characterised by its playful nature. This last element includes the notions of uncertainty of outcome and sanctioned display.

Rules, competition relating to physical prowess and the playfulness inherent in sport make it a unique type of tourism attraction. Specific types of sport, such as football, skiing or BASE jumping, possess their own distinctive traits as tourist attractions but as a whole, they are distinguishable from other broad categories of tourist attractions. As the tourism industry increasingly emphasises experience over products or experience as a product (Tolkach *et al.*, 2016), sport tourism will remain one of its most dynamic components. By analysing sport in the context of these three components, insight can be gained into the way that it functions as a tourist attraction. The impact of changes to the sport attraction can then be considered within the broader context of the spatial and temporal dimensions of sport tourism development.

Scholarship in Sport Tourism

Sport studies and tourism studies share many of the same institutional characteristics. Both are relatively new areas of academia in which academics have worked to establish respected fields of theoretical, methodological and empirical scholarship. Each area has developed systematic interests in sport tourism, which have manifest themselves in a growing body of literature in the form of texts (Gibson, 2006; Higham & Hinch, 2009; Standeven & De Knop, 1999; Weed & Bull, 2009); reviews (Gibson, 1998a; Hinch *et al.*, 2014, 2016; Weed, 2006, 2009); and frequently cited articles (Chalip, 2006; Kaplanidou & Vogt, 2007; Preuss, 2007).

Perhaps the most commonly used argument for focused study in the realm of sport tourism is its economic significance. For example, an analysis of Statistics Canada travel data determined that sport tourism surpassed $6.5 billion (Can) in spending in 2015, an increase of 13% on the previous year (CSTA, 2017). Expenditures by Canadians travelling within their own country accounted for 72% with expenditures by US visitors at 9% and overseas visitors at 18% (CSTA, 2017). In the larger economy in the United States, tourism spending just for amateur sport was estimated at $9.45 billion (US) in 2015, an increase of 5.4% on the previous year (National Association of Sports Commissions, 2017). It is, however, important to recognise that while the revenues associated with sport events in particular may be impressive, the costs can also be high. This is especially true in the context of mega-events such as the Olympics (Zimbalist, 2016). Similarly, it has been argued that social benefits such as community pride can be consciously leveraged in a sport tourism context (Chalip, 2006). Notwithstanding such benefits, there are also potentially negative aspects of sport tourism (Weed, 1999). For example, the introduction of 'nuisance activities' to the countryside, particularly mechanised sports such as trail biking, jet skiing and snowmobiling, has the potential to cause significant negative social and environmental impacts. Adventure and extreme sports, such as BASE jumping, practiced in places like the Lysefjord and Trollveggen regions of Norway, may be associated with safety and liability issues (Mykletun & Vedø, 2002). In such cases, the advantages of developing a better understanding of the dynamics of sport tourism are that these impacts can be recognised, understood and managed in the development process.

Scholarly relevance

The study of tourism and the study of sport are both characterised by numerous hyphenated subfields. In an age of competitive funding in academic institutions, the danger of splintering existing fields of study is a real as well as a perceived threat. Critics have therefore discouraged new subfields. However, the accompanying assumption that insight about the confluence of sport and tourism will emerge organically from within the respective realms of tourism studies and sport studies is problematic. While insights have and will continue to occur independently in each of the respective fields of study, a conscious focus on the intersection is important. It is naïve to assume that interdisciplinary and multidisciplinary approaches will be adopted on anything but an *ad hoc* basis. This situation parallels similar barriers in practice where tourism and sport agencies have often failed to work well together (Weed, 2003). The success of a more focused approach is illustrated by the insights generated by specialised journals such as the *Journal of Sport & Tourism* (Focus Point 2.1). As the theoretical foundations for the study of sport tourism progress (Gibson, 2006), the argument against a specialised field is losing its potency.

Focus Point 2.1: *The Journal of Sport & Tourism*

It is just over 12 years since the reincarnation of the *Journal of Sport Tourism*, an applied/professional journal that was established in 1993, and the emergence of the scholarly *Journal of Sport & Tourism* (*JS&T*) in its place. This change occurred in 2006 and over the course of the last 12 years, the *Journal of Sport & Tourism* has served as the only peer-reviewed academic journal with a specific focus on the intersection of sport and tourism. In this capacity the journal has provided '...not only an important outlet or "cadre" of researchers working in sport and tourism to debate key issues but also an opportunity for authors in related fields to contribute to discussions about sport and tourism, and it helps demonstrate how research in sport and tourism can contribute to wider debates, such as those on climate change and international terrorism' (Weed, 2011: 28). The *JS&T* has become the leading journal for scholars working in this field. It has generated a body of knowledge that forms a foundation that helps scholars to advance the field rather than produce isolated studies that while interesting, fail to advance the field in terms of theory and methodology. Subscription rates doubled in the five years after its repositioning as a scholarly journal which is also reflected in growing impact factors as its articles are increasingly being referenced in other journals. To mark 20 years since the founding of the journal, *JS&T*'s current editor, Professor Mike Weed (Canterbury Christ Church University, UK) identified the 'big questions' that currently face the field in terms of what we know, don't know and should know about sports tourism. Four special issues that address these questions have been planned including two that have been published on sport tourism destinations (Hinch *et al.*, 2016) and theoretical perspectives (Gammon *et al.*, 2017), and two others that are in progress on active sport tourism and spectator events.

Key Reading

Hinch, T.D., Higham, J.E.S. and Moyle, B.D. (2016) Sport tourism and sustainable destinations: Foundations and pathways. *Journal of Sport & Tourism* 20 (3&4), 163–174.

Weed, M. (2011) *The Journal of Sport & Tourism*: A maturing literature. In T.D. Hinch and J.E.S. Higham (eds) *Sport Tourism Development* (2nd edn; pp. 447–450). Bristol: Channel View Publications.

In addition to the parent fields of sport and tourism, there are a multitude of other areas of academic study that overlap and have significant relevance for sport tourism (Figure 2.2). Four examples that are indicative of these many associated fields include events, outdoor recreation, sport management and health & fitness. Event studies include the examination of sports events along with a range of other types of festivals, meetings, conferences and exhibitions (Getz & Page, 2016). While sport events are a popular topic in this field, event researchers seldom highlight the distinguishing features of sporting events relative to other types of events. Even if they did, it is important to recognise that events only comprise one aspect, albeit a high profile one, of sport tourism (see on-site activities in Figure 2.1). In fact, in her analysis of sport travel in Canada, Weighill (2002) found that trips to participate in active sport tourism far outnumbered those associated with trips to spectate at sporting events. As such, it is critical that sport tourism scholarship is expanded beyond the confines of high profile spectator events.

The essence of the outdoor recreation domain is that it tends to occur in natural settings, involving relatively non-institutionalised activities such as canoeing, skiing and surfing. In addition to many popular activities such as hiking, it includes many adventure and extreme sports that often require substantial travel to access unique natural resources. Kane and Zink's (2004) work on package adventure tours, while not explicitly positioned in terms of active sport tourism, provides intriguing insights

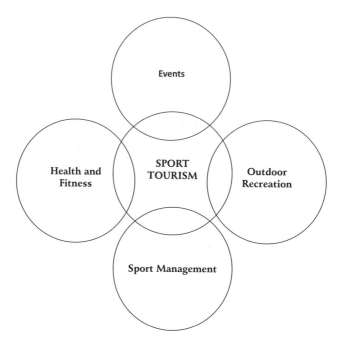

Figure 2.2 Sport tourism and related contextual domains

into this dimension of the field. Once again, there is a clear overlap between outdoor recreation and sport tourism both conceptually and in terms of research activity. However, these domains are not synonymous. A substantial amount of sport activity occurs outside the realm of the natural environment while conversely, many tourism activities that occur in natural settings (e.g. camping and picnicking) are inconsistent with the conceptualisation of sport used in this book.

Sport management tends to focus on highly institutionalised sport. Researchers in this domain typically concentrate on the high performance or development levels of institutionalised sport. The management of sporting events like the Olympics and the FIFA World Cup are popular topics that overlap directly with sport tourism studies (e.g. Sant & Mason, 2015). As in the case of the previous domains, there is a significant shared interest with sport tourism, but sport management scholars tend to ignore non-institutionalised sport travellers and focus on the management dimensions of sport tourism to the exclusion of its other dimensions.

Health and fitness activities represent the fourth domain of relevance that we have highlighted in Figure 2.2. The essence of this domain is evident from both historical and contemporary perspectives. The former is most commonly illustrated by the tourist activity associated with the therapeutic spas of Eastern and Mediterranean Europe in Roman times

(Hall, 1992a). In a contemporary context, travel to partake in therapeutic spas continues with researchers such as Panchal (2014) examining the spa tourists in Asia through the lens of wellness and positive psychology. Many destination spas include sport amenities such as tennis courts and golf courses. While the realm of health and fitness can be defined in ubiquitous terms, it has generally been treated much more narrowly in the literature (Nahrstedt, 2004).

Research in all four of these areas has contributed to the understanding of sport tourism yet the essence of sport extends beyond the collective parameters of these related domains. The defining parameters of sport and their relevance to tourism are not the central interest of research related to events, outdoor recreation, sport management or health tourism. Targeted study on sport tourism can, therefore, provide new and challenging insights that are not normally covered in these associated areas.

The emergence of sport tourism scholarship

Just as the market place has responded to consumer demand for sport tourism products and services, the academic community has responded to a gap in scholarship relating to sport tourism. Initially, this response was relatively isolated and *ad hoc* (Garmise, 1987), but over the past 30 years a growing number of scientific publications has contributed to a maturing body of literature. Gibson (1998a) provided a thorough review of the sport tourism literature as it stood two decades ago. Her critical analysis suggested a maturing body of relevant literature that was still in want of better coordination among agencies at a policy level, multidisciplinary research approaches and greater cooperation between tourism and sport-centred units in academic settings. Seven years later, Weed (2005) acknowledged the growth of publications in this area but used the analogy of a 'brickyard' in arguing that these studies were being grouped randomly in piles rather than being constructed into 'edifices' in which new publications are consciously positioned in terms of proceeding work. Both Weed (2006, 2009) and Gibson (2005) have continued to track the development of the sport tourism body of literature and while noting impressive progress, have called for additional focus on explanation over description. In a meta-analysis of sports tourism research, Weed (2009) identified six substantive areas of progress including insight into: (1) economic impact, (2) leveraging events to obtain desired legacies, (3) holistic approaches to studying impacts, (4) behavioural studies, (5) investigations in destination marketing and image and (6) deeper insight into resident perceptions of sport tourism events.

He concluded that the field has shown signs of increasing maturity as reflected in:

a strong conceptualization of the field; the underpinning of empirical work by appropriate theory; the robust, appropriate and transparent application of methods and methodology; and a clear community of scholars with a sustained interest in the area, served by a credible academic journal and wider body of knowledge. (Weed, 2009: 625)

In their analysis of the sport tourism articles published in five major tourism and sport journals from 2007 to 2011, Hinch *et al.* (2014) found that sport tourism research remained focused on spectator events with an emerging body of research related to a hybrid of the traditional active and event categories (i.e. participant-based events). Almost 60% of the empirically based studies that they reviewed were characterised by quantitative methodological approaches with survey methods being the most common (46%). One of the more telling findings was that the geographic focus of 68% of the articles was North America, Europe and Australia and Oceania with another 13% not having a specific geographic context. The balance, or only 19%, was set in Africa, Asia and South America despite the 2008 Beijing Olympics, the 2010 FIFA World Cup in South Africa, the 2014 FIFA World Cup and 2016 Summer Olympics in Brazil. It has also become apparent that sport tourism scholarship is on the rise in Asia although much of this literature is hidden from Western scholars who do not have the language skills to read articles written in Asian languages (Focus Point 2.2).

Focus Point 2.2: Sport Tourism Research in Japan

Japan is scheduled to host the 2019 Rugby World Cup, the 2020 Tokyo Olympics and Paralympics and the 2021 Kansai World Masters Games. As such, sport tourism is a popular topic in Japan both in terms of practice and scholarship. Given this environment, Hinch and Ito (2018) conducted a systematic review of the English and Japanese literature on sport tourism in Japan from 1990 to 2016. Of these articles, 107 were published in Japanese and 21 were published in English. The vast majority of the articles were published in the last 10 years with 86% of the articles written in English and 64% of the articles written in Japanese being published in 2008 or later. Ninety-one percent of the articles written in Japanese had lead authors at Japanese institutions while slightly less than 50% of the articles published in English had lead authors at Japanese institutions. Fifty-one percent of the articles written in Japanese focused on participation events, 35% on spectator events, 13% on active sport tourism and only 1% on heritage sport tourism. This distribution contrasted significantly with the English language articles, which featured 35% written about participation events, 50% on spectator events, 5% on active sport and 10% on heritage sport. An analysis of the literature addressing the concept of sustainability showed that domestic sport tourists attending participant-based events tended to spend less than general tourists (Nogawa *et al.*, 1996) and questioned the assumptions that there was a substantial economic benefit associated with major spectator events (Manzenreiter, 2008). Several studies examined the social and cultural impacts of sport tourism with one of the key findings being that the bidding process for the Tokyo Olympics fostered a strong link between sport and national identity (Shimizu, 2014). Studies that focused on the environmental impacts of sport tourism were rare but the country's natural resource base was seen as a major asset (Harada, 2016), although it was also noted that some eco-sports (e.g. diving) placed a stress on these resources (Murata, 2010).

Key Reading

Hinch, T. and Ito, E. (2018) Sustainable sport tourism in Japan. *Tourism Planning and Development* 15 (1), 96–101.

The combination of the growing opportunities for employment in the sport tourism field and the growing scholarship in this area has created a fertile ground for sport tourism curriculum development at post-secondary institutions (see Case Study 2.1). Most often, such curriculum is built into existing courses in tourism or sport. Increasingly, however, it has become the focus of independent courses and programmes of study. Such programmes seek a balance between applied knowledge and theory. It is our contention that insight into the growing theoretical foundations of sport tourism will serve students and the field of practice well as graduates will have a better understanding of how and why sport tourism development occurs.

Case Study 2.1: Embedding, Enthusing and Enhancing Employability in Sport Tourism for Students and Graduates

Richard Shipway, Bournemouth University

The coastal towns of Bournemouth and Poole in the south of England, approximately two hours from London, are widely recognised as two of the premier tourism destinations in the UK, and provide a diverse wealth of land- and water-based sport tourism activities. Bournemouth University (BU) has a long tradition of leisure, tourism and hospitality education, and more recently has experienced a rapid expansion of degree programmes that focus on sport and events management. This interaction of the sport, tourism and events industries is embedded within the curriculum of undergraduate and postgraduate programmes at the university with dedicated *Sport Tourism* units taught at both levels to students from BU's tourism, sport, hospitality, leisure and events degree programmes.

(Continued)

Case Study 2.1: (Continued)

To complement more traditional mainstream sports events and activities found at most tourism destinations, the towns have seen the emergence of a plethora of lifestyle-related sport tourism activities. These activities include increasingly popular coastal-based sports such as open water sea swimming events, surfing and standup paddle boarding (SUP). The central hubs of lifestyle sports are the Boscombe Pier area, infamous for the development of the now defunct Bournemouth Surf Reef in 2005, and Bournemouth Pier where adventure and sport tourism activities have been initiated at the Rock Reef Activity Centre and theatre. Reflecting a change in the UK tourism market, more traditional and gentile matinee theatre performances have been replaced by high octane adrenaline-based activities like indoor rock climbing or pier zip wires.

Cycle tourism remains an increasingly popular leisure activity in the region, varying from mountain biking in the Purbeck Hills to road cycling and *sportive* events throughout the New Forest. The rapidly increasing popularity of these activities has led to conflict with local residents due to capacity issues and the contested use of the countryside (Shipway *et al.*, 2016). Bournemouth is also the host venue for the annual Bournemouth Sevens Rugby Tournament, marketed as the *Sporting Glastonbury*. It is hosted over three days, attracts over 30,000 tourists and also incorporates netball, hockey and dodgeball tournaments. Additionally, the towns are in close proximity to professional team sports including the English Premier League (EPL) teams of AFC Bournemouth and Southampton F.C. and the international cricket venue, the Ageas Bowl in Southampton. The increased success and profile of Bournemouth's professional football team has significantly raised global tourism awareness of the town since their promotion to the top tier of English football in 2015. The focus of many sport tourism activities and events is the beach where tourism authorities are proactive in promoting their *seven miles of golden sand*. The towns play host to numerous beach volleyball, rugby, football and handball events which attract tourists from across Europe. On the Sandbanks peninsula, Poole hosts the exclusive Annual British Beach Polo Championship, a combination of high adrenaline polo and evening beach parties.

The content of both the undergraduate and postgraduate sport tourism units at BU explores the foundations and development of sport tourism, before concentrating on the major contemporary issues in sport tourism, various challenges and opportunities, and emerging trends and directions. The increased profile and frequency of major and mega sports events and the globalisation of sport, coupled with the emergence of new sports activities and new emerging destinations, provide numerous case studies to explore. Academic content scrutinises the motivations, behaviours and experiences of sport tourists, and undertakes detailed holistic analyses of the economic, social and environmental impacts of sport tourism that also incorporates both event sport tourism evaluations and leveraging tactics and strategies. Students are encouraged to explore active sport tourism markets including golf, winter, water, lifestyle and adventure sport tourism. The unit draws significantly from theoretical and empirical developments in the *Journal of Sport and Tourism*; associated journals within sport, tourism, events and leisure domains; and where appropriate multidisciplinary perspectives on health and well-being; entrepreneurship and business; or sociology and social psychology, to name but a few areas. The fluctuating and changing global landscape of sport tourism also requires an agile curricular response. An integral part of both sport tourism units are experiential field trips to local sport venues, stadiums and events. Previous trips included visits to the Bournemouth Surf Reef, where students receive guest lectures from local sports tour operators followed by voluntary participation in SUP lessons. Alternative visits include cycling around the New Forest, sailing in Poole harbour or attending cricket events at the Ageas Bowl. These interactive field trips enable students to gain a better understanding of development opportunities and management challenges in sport tourism.

Further along the Jurassic Coast is the borough of Weymouth and Portland, site of the sailing events for the London 2012 Olympic and Paralympic Games. The Games helped raised the profile of Dorset as a tourism destination and left an infrastructure legacy surrounding the sailing venues (Ritchie *et al.*, 2009). During the Games, BU staff and students were involved in operational roles as either official *Games Makers* volunteers, pre-Games torch relay carriers or as seconded employees to the organising committee. These unique, once in a lifetime experiences of staff remain embedded within current curriculum delivery, and for student volunteers the 2012 Games were important for enhancing CVs and increasing employability. Enhancing career opportunities for students is a central aim of the university and many graduating students find themselves working in the sport tourism sector, both during their time of study at BU or through obtaining an initial rung on this career ladder upon graduation. To assist with an agile curriculum response within the UK education sector, BU has a 12-month placement/sandwich year embedded within numerous undergraduate courses, many linked to active sport tourism and sports events.

Recent post-graduation employment by Bournemouth graduates includes jobs in the EPL, for example, a former BSc Leisure Marketing graduate is currently the commercial director at AFC Bournemouth.

(Continued)

Case Study 2.1: (Continued)

BU also has graduates both managing Bournemouth surf schools and working at major golf tour operators, such as Golfbreaks.com. An integral part of the current sport tourism curriculum involves inspirational guest lectures from sport tourism-related industry practitioners. These range from London 2012 venue and stadium managers who highlight legacy benefits and challenges, to senior management from sports marketing and events companies who specialise in professional sailing and outdoor events including running, cycling and winter sports. Inevitably, high profile guest lectures provide insight into the depth and diversity of sport tourism opportunities, from the more traditionally recognised sports to new emerging events and activities.

The key success factors for sport tourism courses at BU include the provision of these experiential field trips to venues or events; creating and developing work experience and employability opportunities within sport, tourism and events industries; a fluid and agile response to reflect constant changes in demand for sport-related activities; and facilitating student engagement with both industry practitioner guest speakers and ex-students currently working within the sport tourism sector. The tripartite concepts of *embedding*, *enthusing* and *enhancing* student employment opportunities have been key success factors for BU's sport tourism courses. Fortunately, the Bournemouth and Poole region provides a *sport tourism laboratory* and destination window for students to critique sport tourism development issues through the programmes developed at BU. These programmes provide theoretical, empirical and practical insights into a diverse phenomenon that offers an equally diverse range of career pathways.

Conclusion

Sport and tourism are closely related in terms of practice. Tourists participate in sports while on their travels and spectators and athletes travel in search of competition or in pursuit of opportunities to engage in their sporting passions. Despite the obvious overlap between sport and tourism, much remains to be discovered about the dynamics of this relationship. Systematic advances in this realm have been greatly aided by focused research in the area of sport tourism rather than the relatively *ad hoc* treatment of the subject prior to 2000. While the study of sport tourism is still relatively young, it has matured into a rigorous scholarly field over the past 25 years. A coherent and insightful body of literature has emerged. But coherence does not imply a single perspective. In fact, a mature area of study embraces multiple perspectives that challenge underlying assumptions and accept and even celebrate the tensions that naturally accompany these perspectives. This depends in large part on the articulation and clarification of these perspectives as attempted in this chapter. Especially important is our conceptualisation of sport tourism as sport-based travel away from the home environment for a limited time where sport is characterised by unique rule sets, competition related to physical prowess and play (Hinch & Higham, 2001). Similarly, we have situated sport as a unique type of tourist attraction. In doing so, we have also adopted an expanded typology of sport tourism that recognises four major types: (1) spectator events; (2) participation events; (3) active sport; and (4) sport heritage. This typology recognises the distinctive nature of participation events, which was formerly lost in the unarticulated overlap between events and active sport. The sport tourism attraction system presented in this chapter attempts to help the reader to consider the distinctive aspects of sport as they relate to tourist experiences. By using this attraction framework and our conceptualisation of sport tourism, it is possible to demonstrate 'what sport tourism shares with, and what distinguishes it from other touristic activities' (Green & Chalip, 1998: 276). These unique aspects of the study of sport tourism are critically explored in the chapters that follow.

3 Sport Tourism Markets

Research into who is a sport tourist, and why sport tourists engage in this sort of tourism may prove to be more complex than is first apparent.
Gibson, 1998a: 57

Introduction

Although it is 20 years since Delpy (1997: 4) commented that a 'travel market focussed entirely on participating or watching sport is a unique and exciting concept', this statement continues to have currency. Sport-related travel is now well recognised as a dynamic and diverse niche sector of the tourism industry (Higham & Hinch, 2009) that may be targeted to broaden the suite of visitor markets that are attracted to a destination (Bull & Weed, 1999). It is a market that may be approached very specifically, in terms of highly specialised competitive sports markets (e.g. Hagen & Boyes, 2016; Moularde & Weaver, 2016) as well as generically, as demonstrated by the mass tourism promotion of major spectator events (e.g. Weed, 2007). The reality, however, is that sport tourism is comprised of a diverse range of niche markets (Collins & Jackson, 2001; Maier & Weber, 1993), which need to be understood through the execution of targeted empirical research (Hinch *et al.*, 2016).

The need to fully recognise the sheer diversity of sport tourism phenomena, and critically understand unique manifestations of sport tourism across a range of spatial scales, is now well established (Higham, 1999). However, it remains the case that the tourism manifestations of high profile, large-scale international sports events are much more prominent in the public and government conscience than unique and specialised regional engagements in sports. In 1999, Bull and Weed (1999: 143) observed that 'sport tourism is really a collection of separate niches but while tourism associated with mega sporting events ... in major urban locations is clearly evident, the potential of sport as a tourism niche elsewhere is perhaps less well appreciated'. While it is not unreasonable to expect tourism destination managers to seek the tourism benefits of large-scale sports events, targeted and critical insights into engagements in niche sports continue to be generally lacking.

Understanding sport tourist markets is an important aspect of the foundation for sport tourism. It is critical that the research community addresses questions that include 'who are sport tourists?', 'what factors motivate sport tourists?', 'to what extent and in what ways do motivations differ between distinct groups of sport tourists?' and 'what travel experiences do sport tourists seek in association with their sporting pursuits?' Addressing these questions in a timely and rigorous manner is necessary, as such insights should inform destination planning and development, resource management and tourism marketing as they relate to the sustainable management of sport resources and tourism destinations (Hinch *et al.*, 2016). Market analysis, then, is critical to the effective development of sport tourism within the context of regional or national tourism destinations. The first part of this chapter discusses conceptual approaches to the classification of sport tourist types. This is followed by an examination of sport tourism markets and the means by which these markets are effectively understood, and a consideration of the challenges and opportunities arising from insights into different forms of engagement in sport tourism.

Conceptualising Demand for Sport Tourism

Conceptualising sport tourism is a useful starting point in the study of sport tourism markets. The distinction between participation and spectatorship, for example, is important, although engagements with sport have evolved and merged (Lamont *et al.*, 2014). Glyptis (1989, 1991) deployed the terms 'general dabbler' and 'specialist' to describe different levels of tourist engagement in participant and spectator sports. Hall (1992b) also identified two types of active sport tourists: 'activity participants' who regard their participation as a medium of self-expression and 'players' who are competitive in their engagements in sport. These terminologies offer insights into different drivers of engagement in sport, although such engagements have expanded to include, for example, those who take part in serious leisure (e.g. competitive amateur athletes), negotiating various constraints in order to do so (Kennelly *et al.*, 2013). The distinction between 'sport-orientated holidays' and 'less sport-orientated holidays' has been the conceptual basis for the study of sports activities that serve as important tourist motivations (World Tourism Organisation & International Olympic Committee, 2001). Various typologies have been developed over time, in an attempt to describe the broad range of engagements in sport and tourism (Gammon & Robinson, 1997; Maier & Weber, 1993; Reeves, 2000; Standeven & De Knop, 1999).

Clearly, tourists who engage in sports at a destination do so with varying degrees of commitment, competitiveness and active/passive engagement (Gibson, 1998a). The sport tourism market may be segmented on these grounds into niche markets or 'demand groups', which differ in many aspects of the visitor experience (Chapter 8). Maier and Weber (1993) identified four demand groups based on the intensity of the sports

Table 3.1 Sport tourism demand groups and requisite visitor facilities

Demand groups	Visitor demands and required facilities
Top performance athletes	Efficiency is the priority. Access to competition and suitable training conditions and facilities are the priority for these travellers. When meeting the priorities of this group, tour organisers and destination managers need to give consideration to specific accommodation and dining demands (e.g. dietary requirements) as well as access to physicians, injury rehabilitation facilities and other performance-related services.
Mass sports	Preserving health and maintaining fitness is the aim of this demand group. Performance targets are individually fixed. The accessibility of holiday regions and the quality of sports facilities are the key considerations for this market segment.
Occasional sports (wo)men	Compensation and prestige play greater roles than sporting ambition in the pursuit of occasional sports. This demand group gives preference to less demanding sports such as recreational skiing and bowling. Sporting activities receive no greater priority over cultural sightseeing and other interests within this market group.
Passive sports tourists	No individual sports activities are pursued. The focus of this group lies with mega sports events and distinguished sports sites. Includes coaches and attendants to high performance athletes, as well as media reporters. Requires high volume infrastructures to accommodate the needs of large numbers of event sport attendees.

Source: Maier and Weber (1993: 38).

activities that are pursued at a destination (Table 3.1). They then used these demand groups to usefully describe the unique resource development requirements for each. For example, the resource requirements that top performance athletes seek at a tourism destination relate specifically to the enhancement of performance in sport (e.g. training, sports science and sports medicine facilities).

Much less is known about the importance of leisure and unique tourism experiences at places of training or competition in terms of how they may serve the interests of athletes to achieve 'tour balance' (Higham & Hinch, 2009; Hodge et al., 2008). 'Tour balance' describes the need for elite athletes to occasionally escape from routine (e.g. training schedules) and the high pressures of performance. Opportunities to experience the stimulation of new and unique places are now recognised as an important aspect of sports performance through periodic relief from the pressures of competition (Hodge & Hermansson, 2007). Hodge et al. (2008) consider 'tour balance' in professional sport to be critical not only to achieving optimum performance at the place of competition, but also to countering the prospects of 'burn out' in the longer (career) temporal dimension. 'Occasional sports (wo)men' and 'passive sport tourists' stand in significant contrast, given the priority that they are likely to place in the experience of place and tourist experiences that a destination may offer. These visitors are more likely to be motivated by the 'culture of sport' and the heritage values that may be associated with sports attractions (Gammon & Ramshaw, 2013).

In a similar vein, and building upon the earlier contribution of Maier and Weber (1993), Reeves (2000) identified six 'types' of sport tourists and explains the distinctions between them in terms of decision-making, motivations, lifestyle and spending profiles (Table 3.2). The diversity that exists within the generic sport tourism market is, once again, highlighted. However, it should be noted that Reeves's (2000) typology is illustrative rather than definitive. It contains generalisations relating, for instance, to visitor expenditures that require the support of rigorous research. Autonomy in destination choice may indeed be limited for 'driven' athletes who may be subject to structured training programmes and schedules in advance of competition (i.e. professional sports organisations), particularly in the case of team sports. However, in many cases, the preferences of 'driven' (elite) athletes are now accommodated through negotiated arrangements (usually mediated by management agencies, managers or legal advisors) that may extend to periods of sabbatical leave, and a degree of autonomy over travel, training and competition schedules.

Robinson and Gammon (2003) advance the conceptualisation of sport tourists based on the motivations held by tourists vis-à-vis their involvement in sport (Table 3.3). Their contribution lies in the distinction between two forms of sport tourism based on the primacy of tourist motivation. They use the term 'sport tourism' where sport is the primary travel motivation with other touristic activities an important but secondary element of the tourist experience. Alternatively, in the case of 'tourism sport', sport serves as a secondary or incidental component of the tourist experience. Gammon and Robinson (1997) also distinguish between active and passive involvement in competitive and non-competitive sports. Therefore, both 'sport tourism' and 'tourism sport' may be defined in terms of hard and soft participation. The distinction between the two lies in the seriousness with which tourists engage in their chosen sports. This conceptual framework captures the diversity of the sport tourism travel market, which varies along scales of participation and competitiveness, and where sport may function as a primary, secondary or purely incidental travel motivation. It also complements the concept of the tourism attraction hierarchy (see Chapter 2).

Gammon and Robinson (1997), therefore, identify three dimensions that highlight the variation that exists within the demand-side of sport tourism. They include the status of the sport activity in the motivational profile of the tourist (primary, secondary or incidental), the type of involvement in the sport activity (active or passive) and the competitive or non-competitive nature of the sport activity. In doing so, they contribute to a better understanding of sport tourism consumer markets, which provides insights into the distinct sport and tourism-related services and experiences required by each. It is the extent to which engagement in sport is pursued by tourists that also forms the basis of Standeven and De Knop's (1999)

Table 3.2 Sport tourism types and visitor profiles

Type	Decision-making	Participation	Non-participation	Group profile	Lifestyle	Spending
Incidental	Unimportant	Out of duty	Not relaxing, holiday-like	Family	Sport is significant	Minimal
Sporadic	Relatively important	If convenient	Easily contained/ put off	Friends and family	Non-essential	Minimal except for 'one-offs'
Occasional	Sometimes determining	Welcome addition to tourist experience	Other commitments	Often individual especially business tourists	Conspicuous consumption	High on occasions
Regular	Important	Significant part of enjoyment	Money or time become prohibitive	Group or individual	Important	Considerable
Dedicated	Very important	Central to experience	Due to unforeseen barriers	Individuals and groups of like-minded	Defining element	Extremely high and consistent
Driven	Very important but little autonomy	Sole reason	Through injury or fear of it	Elite groups or solitary	The profession	Extremely high but funded by others

Source: Reeves (2000).

Table 3.3 Conceptualisation of sport tourists based on sport and travel motivations

Sport tourism:	Individuals and/or groups of people who actively or passively participate in competitive or recreational sport while travelling. Sport is the prime motivation to travel although the touristic element may reinforce the overall experience.
Hard definition	Active or passive participation in a competitive sporting event. Sport is the prime motivational reason for travel (e.g. Olympic Games, Wimbledon or the London Marathon).
Soft definition	Active recreational participation in a sporting/leisure interest (e.g. skiing, walking, hiking, kayaking).
Tourism sport:	Active or passive participation in competitive or recreational sport as a secondary activity. The holiday or visit, rather than the sport, is the prime travel motivation.
Hard definition	Competitive or non-competitive sports act as an important secondary motivation that enriches the travel experience (e.g. sports cruises, health and fitness clubs).
Soft definition	Competitive or non-competitive sport or leisure as a purely incidental element of the holiday experience (e.g. mini golf, indoor bowls, ice skating, squash).

Source: Gammon and Robinson (1997: 10–11).

conceptual classification (Table 3.4). The sport activity holiday segment of Standeven and De Knop's classification illustrates the diversity of the sport tourism market. Single sport activity holidays accommodate those tourists who seek to engage in specific sports, such as downhill skiing, cross-country skiing or snowboarding. The wider touristic element of the destination may carry little sway in the travel decision process in these instances. This market stands in contrast to the multiple sport activity holidays market for whom opportunities to engage in sporting pursuits are more broad ranging, casual and less likely to be the sole focus of visitor activity.

Table 3.4 Sport tourism activities classification

Classification	Examples
Sport activity holidays	
o Single sport activity holidays	Skiing, cycling, trekking
o Multiple sport activity holidays	Sports camps, holiday clubs (e.g. Club Méditerranée)
Holiday sport activities	
o Organised holiday sport activities	Golf, rafting, cruise ship sport activities
o Independent holiday sport activities	Adventure activities (e.g. bungee jumping)
Passive sports on holidays	
o Connoisseur observer	Olympic Games, Masters Golf, Wimbledon Tennis Championship, Kentucky Derby, museums, halls of fame, stadium tours
o Casual observers	Hurling (Ireland), Thai boxing (Thailand), bull fighting (Spain)
Active sports during non-holiday time	Training camps, recreational sport during business and conference travel
Passive sports during non-holiday time	Dragon boat racing spectatorship while in Hong Kong on business

Source: Standeven and De Knop (1999: 88).

These, and other segments within this classification, are associated with distinct market characteristics. The terms 'active' and 'passive' are used in this classification in reference to sport participation and non-participation, respectively, although this dividing line is increasingly blurred. These should not be confused with other forms of active engagement in sport tourism such as team management or support crews for those who engage in competition. Furthermore, in some sports, spectators are encouraged to be active in their support of a competitor or team (e.g. banner competitions and inter-session spectator competitions) as a means of generating atmosphere at stadiums and other sports venues. In other cases, spectators need no such encouragement in their expressions of contemporary culture, which may relegate the sports display to secondary importance in relation to performances of fan identification.

These typologies, which were developed and published between 1993 and 2000, serve an important role in the academic development of the field of sport tourism. They provide industry, government and academics with important insights into the range of tourist types that are represented within the sport tourism market. They also provide marketing professionals with insights into distinct manifestations of sport-related tourism, by way of clearly conceptualising the range of sport tourist types. However, while offering important contributions to the development of this field of scholarship, these typologies have been critiqued as limited in that they may oversimplify complex phenomena (Weed, 2005). It has been argued that the typological approach, while useful in terms of *explanation*, is insufficient to advance an *understanding* of sport tourism phenomena (Weed, 2005).

Thus, since 2005, the academic community has attempted to move the field of sport tourism forward from description and explanation to critical understanding (Gibson, 2006; Higham & Hinch, 2009; Weed, 2005). In doing so, Weed (2005) argues against the need to establish the primacy of either sport or tourism, suggesting that 'the primacy of the sport or the tourism element in many sports tourism experiences cannot be established and, in fact, for many experiences separate and distinguishable sport and tourism elements may not be present' (Weed, 2009: 619). Rather, it is the unique cultures of different places, which extend to the cultures – values, beliefs, behaviours (including styles of play and interpretations of rules) and traditions – of sport (Hinch & Higham, 2004), that contributes to a complexity in the study of sport tourism that challenges typological classification. Instead, Weed (2005) proposes that all sport tourism phenomena represent a complex and unique interplay of activity (diverse engagements in sport), people (hosts and guests) and place (unique locations where sports occur) (Figure 3.1). This conceptualisation has usefully driven the development of the field towards understanding unique

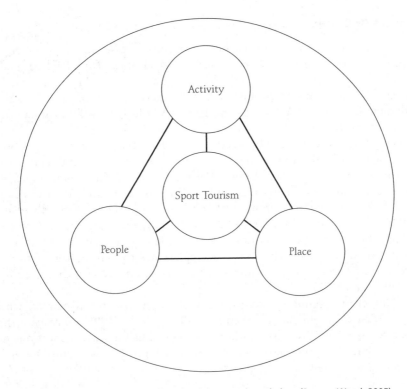

Figure 3.1 Sport tourism as an interplay of activity, people and place (Source: Weed, 2005)

sport tourism phenomena in relation to the complex interplay of activity, people and place (e.g. Evans & Norcliffe, 2016; Morgan, 2007).

The systematic treatment of sport tourism markets has been well served by Redmond's (1990) tripartite sport tourism classification, which includes sport vacations, multi-sport festivals and world championships, and sports halls of fame and museums. Gibson's (1998a) literature-based analysis of sport tourism presents a variation of Redmond's classification. Her classification includes active, event and nostalgia sport tourism. Here, we build upon the earlier contributions of Gibson (1998a) and Redmond (1990) by proposing a fourfold classification of sport tourism:

(1) Spectator events.
(2) Participation events.
(3) Active engagement in recreation sports.
(4) Sports heritage and nostalgia.

This schema serves as the framework for the following discussion.

Spectator Events

Event sport tourism, in its most prominent guise, involves travel to experience sporting events, where the body of spectators outweighs a small number of elite competitors (Hinch & Higham, 2004). The most widely researched examples of spectator events focus on mega-events such as the Olympic Games, FIFA World Cup, Rugby World Cup and Formula One Grand Prix (Burgan & Mules, 1992; Fourie & Santana-Gallego, 2011; Jones, 2001; Ritchie, 1984; Weed, 2007). However, sports competition may or may not be the primary attraction of a sports event. Hallmark sports events such as the Wimbledon Lawn Tennis Championship may also be attended for its heritage and traditional value, and Americas Cup and the Super Bowl for commercial and business purposes (O'Reilly *et al.*, 2008). Large-scale spectator sports events may, then, attract tourists for whom the sporting competition is a coincidental or secondary factor in their attendance. This suggests that approaches to market segmentation are also applicable to spectator event markets, an exercise that requires further empirical research.

Numerous avenues of tourism development may be associated with spectator-driven elite sports events. Faulkner *et al.* (1998) emphasise the need for sport and tourism authorities to establish a set of conditions to ensure that this potential is captured. In a study of the 2000 Olympic Games, they state that

> ...in reality, there are both tourism opportunities that can be derived from hosting the games and offsetting negative effects, and the degree to which the former are accentuated and the latter ameliorated ultimately depends on the extent to which the leveraging strategies adopted by the industry and relevant public agencies are effectively integrated. (Faulkner *et al.*, 1998: 1)

The build-up to the Sydney 2000 Olympic Games involved a coordinated leveraging programme to which both sport and tourism administrators at the federal and state levels were required to actively contribute. The outcome was effective destination promotion, successful pre-games training and acclimatisation camps, the stimulation of convention and incentive travel, the promotion of pre- and post-Games travel itineraries and the minimisation of diversion and aversion effects (Faulkner *et al.*, 1998; O'Brien, 2006). Leveraging elite sports events requires a clear understanding of the tourism development opportunities that exist beyond games-induced travel. Indeed, various modern Olympic Games have been studied to forecast and/or measure Games-induced travel (Kang & Perdue, 1994; O'Brien, 2006; Pyo *et al.*, 1991). It is important to note that the scope and scale of Olympic tourism may be easily overestimated and oversold

(Weed, 2007) and, as Faulkner *et al.* (1998: 10) observe: 'once the normal level of travel to the host city is factored in to estimates the net effect of games induced travel is much reduced'.

It is important to acknowledge the complex flows of visitors into and out of a spatial unit of analysis (which varies with the defined scale of the event destination) (Preuss, 2007; Weed, 2007). The diversion effects (real or perceived) may be encouraged by media attention paid to capacity constraints at the destination (e.g. traffic congestion, over demand for accommodation, resulting in inconvenience, inflated costs and security issues), which may be placed under strain when events take place. Pyo *et al.* (1991) identify a range of factors that may discourage attendance at summer Olympic Games including political boycotting, price gouging, crowding and congestion, and security concerns. Ticket distribution may also bear upon the propensity of sport tourists to attend events such as the Olympic Games (Thamnopoulos & Gargialianos, 2002). Chalip *et al.* (1998) present an analysis of sources of interest in travel to attend the Olympic Games. Such analyses provide useful insights into the relative importance of the event itself, and the destination hosting the event, in terms of the travel decision-making process. These studies offer information that is critical to the successful leveraging of sports events.

Sport tourism markets vary from one event to another. However, useful generalisations are possible. For example, Faulkner *et al.* (1998) employ the term 'sport junkies' to describe tourists who visit a destination specifically to attend a sporting event, but demonstrate little propensity to engage in pre- and post-event itineraries. This term describes a sport tourism market that is single-mindedly focused on the sports event itself. Chalip (2001) observes that many Australian cities targeted specific market segments during the lead up to the 2000 Olympic Games. These efforts often targeted specific national Olympics teams and the travelling fans that follow their teams.

Similarly, the Germany 2006 FIFA World Cup fostered links between national teams competing in the World Cup, and municipalities that hosted those teams during the tournament. This strategy allowed the development of links between specific football teams, the host cities and the spatial mobilities associated with the large groups of travelling football fans that wanted to be located in close proximity to 'their' teams. There is increasing evidence that the behaviours of travelling fans may be influenced by the location (e.g. training bases) of 'their' teams in advance of and during competition. The 'Barmy Army' is a case in point. This is a 'group of England cricket fans who spend vast amounts of time and money travelling the globe to watch what has been in recent years a less than successful team' (Weed & Bull, 2012: 127). Their travel decisions are closely tied to the tour itineraries of their team.

In contrast to 'sport junkies', sport spectators may engage in a more casual relationship with sport. In these cases, 'general interest and match attendance can shift in response to wins and losses, the state of the venue, the appearance of star players and a change in the weather' (Stewart, 2001: 17). Casual consumers of sport present sport markets with unique challenges associated with accessing this element of sport tourism demand. The importance of understanding the travel motivations and wider pre- and post-event itineraries of sports tourists emerges clearly from these studies. The leveraging of sport tourism events requires consideration of the sport and tourism product at the destination; the supply and demand for sports facilities and services; and tourism experiences before, during and after the event (Faulkner et al., 1998). Much less is known about the tourism experiences of elite athletes. Conceiving elite athletes as tourists who visit destinations, albeit for different reasons to those who travel to destinations to experience their sporting performances, is an interesting proposition. Some destinations now seek to actively attract and cater for elite athletes and sports teams (Chalip, 2004b; Francis & Murphy, 2005).

Participation Events

Much research in the field of event tourism focuses on large-scale spectator events (Weed, 2007). However, this is only a partial picture of event sport tourism (Bull & Weed, 1999; Gratton et al., 2005). Event sport tourism includes competitive non-elite, amateur and recreational sports events, where the number of competitors may be large, and the number of spectators negligible or non-existent (Figure 3.2). Exceptions to this general rule do exist where non-elite events attract large numbers of family and friends as spectators (Carmichael & Murphy, 1996). In some instances, elite and non-elite competitors are accommodated in a single event, which creates a broad catchment of elite athletes (and spectators) and non-elite competitors. The London, New York and Boston marathons provide evidence of the success of such events. The relationship between participation and spectatorship in event sport tourism deserves more academic attention. These distinct forms of event sport tourism justify separate analysis as the markets, promotional possibilities, infrastructure requirements, tourist behaviours, travel patterns and associated tourist experiences of each are likely to stand in significant contrast.

Bale (1989: 114) observes that 'even quite small sporting events can generate substantial amounts of revenue for the communities within which they are located'. The trends outlined in Chapter 1 help to explain why participation sports have experienced such growth in recent years, which creates leveraging opportunities for tourism destinations (Derom & Ramshaw, 2016). Destinations that are unable to host large-scale sporting

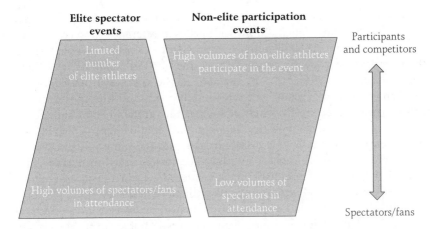

Figure 3.2 Conceptualisation of the relative engagement of spectators and participants in elite and non-elite sport events

events due to capacity constraints may compete to attract sub-elite or non-elite participation-based sport events (Gratton *et al.*, 2005; Higham, 1999). Chogahara and Yamaguchi (1998), for example, report on the National Sports Festival for the Elderly (Japan). This study identifies that participants tend to engage in a wide range of tourist activities, particularly sightseeing and visiting hot spas both during and following the completion of the event. This study confirms that those who participate in lesser or non-competitive sports events are more likely to take advantage of opportunities to engage in tourist activities at a destination. Little is known about the travel preferences of participants who form the less competitive element of the event sport tourism market. This niche market is characterised by travel motivations that are distinct from elite and competitive event participants.

The distinction between elite and non-elite event sport tourism is important (Figure 3.2). Carmichael and Murphy (1996) suggest that event sport tourists should be differentiated on the basis of spectatorship and participation, the latter including athletes, officials and coaches. Their study provides insights into visitor origin, length of stay, expenditure patterns, numbers of friends and relatives accompanying participants, and their intentions to return to the town hosting the event. This points towards the tourist motivations that distinguish those who attend elite sports events and those who travel to participate in non-elite sports competitions (Chapter 8).

In response to this point, some attention has been usefully paid to 'second-order' (sub-elite) sports events which have been pursued by second tier or provincial cities seeking to 'shine on the global stage' (Whitson, 2004: 1221). Gratton *et al.* (2005) highlight the sharply contrasting tourism contexts associated with second tier sports events which serves as a warning to guard against the overselling of the tourism and economic benefits of hosting such events. Black (2008: 467) notes in reference to 'second-order' sports events that 'their benefits continue to be chronically oversold and their opportunity costs minimized or overlooked'. Similarly, Whitson and MacIntosh (1996: 288) observe that in the absence of an established tourism trade '…it is a mirage to think that a substantial tourist economy can be constructed on the back of such events alone. The imaging effect is too small and the competition is too great'. While Higham (1999) notes the potential for small-scale events to generate the same positive impacts as larger events (within their own geographical scales of analysis), it is critical that these are realistically considered within the infrastructural and 'tourist economy' limitations of smaller communities.

These points aside, one of the key justifications for the separate treatment of participation sports events in this book is the phenomenal growth in participation in highly competitive, sub-elite sports events in recent years. This has been a feature of the evolution of sport tourism since the publication of the previous edition of this book in 2010. Many such participants are engaged in serious training schedules in preparation for their target events, but are self-referenced, as opposed to results orientated (Falcous, 2017) in their goals. Such events include sports such as mountain biking, triathlon and Ironman, which involve many participants who train, compete and measure their performance against themselves rather than against other competitors.

Active Engagement in Recreational Sport

The active sport tourism market is constituted of individuals who pursue physical involvement in competitive or non-competitive sports while travelling. These engagements may be defined as 'de-sportified' in that they are 'loosely structured, non-competitive, and social connective' (Falcous, 2017: 1). A number of published articles examine the active sport tourism market (Funk & Bruun, 2007; Getz & McConnell, 2011; Gillett & Kelly, 2006; Green & Chalip, 1998; Shipway & Jones, 2007; Yusof & Douvis, 2001). However, Gibson (1998a: 53) notes that active sport tourism market research is generally 'scarce, usually descriptive and typically atheoretical'. In 1992, Yiannakis and Gibson introduced the term 'sportlover' to describe the growing travel market represented by individuals who are physically active and prefer to remain so while traveling for business and/or leisure. This includes the established markets of tourists who pursue sport activity holidays (e.g. skiing and snowboarding), and those who engage recreationally in sports activities while on holiday (Standeven & De Knop, 1999). More significant, perhaps, is the growth in tourism phenomena associated with those who are motivated to travel to specific locations to actively engage in their sporting passions. These active participants may engage in sport to develop their specific skills and abilities or seek competition, but increasingly these visitors set out to experience first-hand unique or famous sports places, perhaps in order to develop a sense of personal or collective identity, and to develop their standing within a sport subculture (Green & Chalip, 1998; Higham & Hinch, 2009). These forms of active sport tourism may also allow participants to engage with the elites of their sports.

The active sport tourism market has typically been portrayed as physically active, college educated, relatively affluent and 18–44 years old (Delpy, 1998), but the assumption that active participants in sport can be generalised in any way masks a wide and growing diversity within this broad classification. A failure to recognise and capture the diversity of the broad active sport tourism market does the field a disservice (Green & Chalip, 1998; Lamont *et al.*, 2014; Shipway & Jones, 2007). In fact, both those who engage in sport (Ross, 2007; Stewart *et al.*, 2003; Taks & Scheerder, 2006) and tourism (Bieger & Laesser, 2002; Dolnicar, 2002) are subject to various approaches to market segmentation. Similarly, various approaches to segmentation have been applied to the active sport tourism market (e.g. Gibson *et al.*, 1998).

Geographic market segmentation

Geographic segmentation of the active sport tourism market, based on visitor origins or market location, is a popular approach to the practice of sport tourism. The geography of sport establishes the link between place of residence and exposure to specific built or natural (e.g. seasonal) sports resources and, therefore, opportunities to engage in certain sports in specified locations (Bale, 1989; Rooney & Pillbury, 1992). Proximity to sports resources, be they natural (e.g. surf beaches), built (e.g. sports stadia) or a combination of the two (e.g. ski resorts), bears upon the propensity to consume certain sports as competitors, participants and/or spectators. Place of residence also exposes individuals and communities to the sport cultures of a place, which although challenged by the relentless progress of global forces in sport and tourism (Higham & Hinch, 2009), continue to exist at a range of spatial scales of analysis (see Focus Point 3.1). It has been argued that the safeguarding and embellishment of unique regional sports cultures are critical to the uniqueness of sport and, therefore, the sustainability of sport-related tourism (Hinch & Higham, 2005).

Focus Point 3.1: Sport and Space

More than a game, sport is 'a social phenomenon in its own right' (Laidlaw, 2010: 49). Indeed, Laidlaw (2010) explains that sport matters so much to national identity that it can alter the course of social and political history. Sport functions as a social phenomenon at a range of spatial scales. Laidlaw (2010) notes the important role of sports

(Continued)

Focus Point 3.1: (Continued)

in forging national identity in countries such as New Zealand (rugby union) and Australia (rugby league) just as in the cases of India (cricket), Brazil (football), Canada (ice hockey) and Ireland (Gaelic football). The 'entangling of sport and national images is common' (Ward, 2009: 521). Similarly at a regional level, the north of England developed as a stronghold of rugby league in response to the exclusivity of rugby union in the English Home Counties (Laidlaw, 2010). Rugby union in North America demonstrates equally entrenched patterns of participation; both in the United States and Canada the sport prevails on the West Coast. The same even applies at urban and suburban levels of analysis. Despite attempts to globalise sports, Melbourne (Victoria) remains an Australian Rules stronghold, as Sydney (New South Wales) remains the Australian home of rugby union and Brisbane (Queensland) rugby league. The same sociological patterns can even be discerned in suburbia. Sydney working-class suburbs, such as Balmain and Paramatta, are loyal rugby league communities. In Wellington (New Zealand), the suburb of Wainouiamata is defined by rugby league as Petone is by rugby union (Laidlaw, 2010). As such, identification with sports clearly takes place at national, regional, local and even suburban scales of analysis.

Key Reading

Laidlaw, C. (2010) *Somebody Stole My Game*. New York: Hachette.

Understandably then, the link between nationality and patterns of sport participation is well established (Breuer *et al.*, 2011; Wicker *et al.*, 2013). Patterns of active sport tourism participation can be generalised at a national level in terms of participation rates, engagement in specific sports and gender and participation (Coakley, 2017). In Canada, for example, participation in sports is highly concentrated in relatively few sports: golf, ice hockey, swimming, soccer, basketball, baseball and skiing; and marked by a significant male gender imbalance in participation rates (Ifedi, 2008). Studies such as these provide sport and tourism managers with rich information relating to the increasing intersection between and the changing patterns of sport participation and travel preferences.

Socio-economic market segmentation

Socio-economic market segmentation is based upon variables such as occupation and income (Swarbrooke & Horner, 1999). Participation in inexpensive, team-based contact sports like street basketball and baseball is typical of lower socio-economic urban youth in North America and Cuba (Thomson, 2000). So too are low barrier, low-cost individual sports such as basketball, boxing and mixed martial arts (MMA), which have paved the way for some athletes to break cycles of disadvantage, abuse or poverty (Coakley, 2017). By contrast, expensive, individual and non-contact sports have long been favoured by the upper social classes (Yiannakis, 1975). Booth and Loy (1999: 10) argue that 'sports such as golf, tennis, sailing, show jumping and skiing... reflect the upper class's unique aesthetic and ethical dimensions, temporal/spatial orientations, material and symbolic status signs...'. These consumers of sport are '... free to play sport at midday, mid-week or out of season [by travelling to the opposite hemisphere], and ... have the resources to play in exclusive and secluded places: cloisters, country clubs and lodges, and private game reserves' (Booth & Loy, 1999: 11). Sport may be used to reproduce and entrench class distinctions, in exactly the way that distance is used to the same effect in tourist holiday decision-making (Casey, 2010).

The links between sport and socio-economic status, while also subject to the forces of social and economic globalisation (Higham & Hinch, 2009; Maguire, 2000), are inescapable, both in terms of participation and spectatorship. Many sports in England – croquet, polo, tennis and rugby union – according to Laidlaw (2010: 49) have 'served to divide for a very long time. The exclusive preserve of the English public schools... unimaginable to those beneath the upper middle class'. The masses, Laidlaw points out, no doubt attracted by the simplicity and accessibility of the sport, took to football instead, both in England and around the world. Rugby union remains, in terms of spectatorship, the domain of the gentile in contrast to the local tribal hordes of the fandoms associated with football clubs across countries and continents around the globe. While democratisation has been the dominant trend in recent decades (Coakley, 2017; Standeven & De Knop, 1999), these sports entrench and reaffirm socio-economic standing (Booth & Loy, 1999).

Demographic market segmentation

Swarbrooke and Horner (1999: 95) confirm that 'segmentation based on subdividing the population on the basis of demographic factors has proved particularly popular in tourism'. The demographic profiling of sport tourism markets in North America shows, for instance, that active participation in sports varies on the basis age (Ifedi, 2008; Loverseed, 2001). The most popular participation sports in the United States include recreational swimming (94%), recreational walking (83%) and bowling (74%). Activities such as fitness walking, treadmill exercises and stretching, and sports such as golf and fishing, are the preference of the seniors market (55 years and older), while basketball, soccer and baseball are favoured by the youth market (6–17 years). Participation in golf is influenced by demographic variables such as income, senior citizenry and 'empty nest' status (Tassiopoulos & Haydam, 2008). In Canada, participation in sports such as ice hockey, volleyball and football is dominated by affluent, young, males who are typically students (Ifedi, 2008). Hudson *et al.* (2010) demonstrate that the perceptions of constraints acting upon participants in downhill skiing vary on the basis of cultural or ethnic background (distinguishing in their study between Chinese and Anglo-Canadians) in Canada.

Gibson *et al.* (1998) provide one of the more detailed demographic analyses of the active sport tourism market from a life-span perspective. They profile the active sport tourist market in the early adulthood (17–39 years), middle adulthood (40–59 years) and late adulthood (60–91 years) life stages. While active sport tourism proved to be pursued particularly by those in early adulthood, 'a sizeable number of both men and women choose sport orientated vacations in middle and late adulthood as well' (Gibson *et al.*, 1998: 52). They, like Harahousou (1999) and Tokarski (1993), identify physical activity among people in late adulthood as an increasing trend. The active sport tourism market is also influenced by changes in societal conventions regarding female participation in sports. Gibson *et al.* (1998: 54) state that 'the subject of gender and sport is full of examples showing how gender-typed social expectations affect women's participation in sport and physical activity'. This situation is emerging from historical male domination of participation, particularly in contact sports, due to changing societal ideologies about the gender-appropriateness of many activities (Carle & Nauright, 1999; Wiley *et al.*, 2000; Wright & Clarke, 1999). Women's rugby union and football have become major growth markets, which, in turn, has stimulated the study of differentiated sport participation experiences on the basis of gender (Green & Chalip, 1998).

Psychographic market segmentation

Psychographic studies are founded on the premise that 'the lifestyle, attitudes, opinions and personality of people determine their behaviour as consumers' (Swarbrooke & Horner, 1999: 96). For example, the psychographic profile of *sports-for-all* participants differs from those who pursue technical challenge or competition through active involvement in sports. The defining criteria of sport-for-all include the absence of entry qualifications, championship prizes and competition between participants (Nogawa *et al.*, 1996). Instead 'sport-for-all ...emphasises the joy of sport participation and health-related fitness while de-emphasising excessive competition. The concept of a sport-for-all event is that every participant is a winner' (Nogawa *et al.*, 1996: 47). The active sport tourism market can be effectively segmented based on the differences between sport-for-all participants and their more competitive counterparts.

Active participation in some sports may also be associated with distinctive subcultures that are an expression of identity (Green & Chalip, 1998). Wheaton (2000, 2004) examined one such subculture in her ethnographic study of windsurfing. She observed that the emergence of new and individualised leisure sports, such as windsurfing, snowboarding and mountain biking, represents 'much more than …intermittent recreation; participants are involved in a multi-layered leisure subculture' (Wheaton, 2000: 256). Subcultures are expressed in various ways, including life choices such as career, work time, place of residence and tourist destination preferences (Wheaton, 2013). Similar conclusions have been made in the case of snowboarding (Heino, 2000). Both studies suggest that the values associated with individual sports can be instrumental in shaping the attitudes and personalities of participants (Chapter 6). Less organised and regulated sports, such as beach volleyball, street basketball, mountain biking and skateboarding, 'emphasise values such as excitement, spontaneity, rebellion, non-conformity, sociability and creativity, and these are assuming considerable importance within the context of youth culture' (Thomson, 2000: 34).

The psychographic profiles of sport tourists are also shaped by culture. Cross-cultural studies have shown that norms for experiencing emotions differ between cultures, with implications for behaviour, well-being and happiness (Eid & Diener, 2001). This approach has been effectively used to analyse generic differences between European, American and East Asian cultures in terms of propensity towards high-arousal positive affect states (e.g. active engagement, excitement and adrenaline) and low-arousal positive affect states (e.g. passive engagement, calm and peacefulness) (Tsai, 2007) (see Case Study 3.1). Of course, variation exists within cultures, and engagements in sport may evolve over time. Japanese visitors to New Zealand have been typically associated with relatively passive activities such as viewing scenery, hot air ballooning and golf. This market has evolved over time to become more engaged with activities such as skydiving, white-water rafting, jetboating and adventure caving (Pavlovich, 2003). These dynamic trends in turn contribute to the evolution of tourism destinations (Higham, 2005).

Similarly, it has been argued that career stages unfold over time in association with the development of subcultural identities, which may involve travel to specific events or particular locations as part of the career ladder (Getz & McConnell, 2014). Such studies 'highlight the utility of leveraging event consumers' identification with a sport's subculture when promoting sports events' (Green & Chalip, 1998: 288). Indeed, Green and Chalip (1998) propose that active participation in sport at a destination may give priority to sharing and affirming their identities over the competitive element of participation.

Case Study 3.1: Culture, Ideal Affect and Sport Tourist Motivation

Eiji Ito, Faculty of Tourism, Wakayama University, Japan

Affect valuation theory (Tsai *et al.*, 2006) has been employed to investigate the associations among leisure, emotions and culture. Tsai (2007) proposed that leisure activities might be a way people try to reduce the discrepancy between their ideal and actual affect. More specifically, people who want to feel high-arousal positive affect (i.e. their ideal affect) would like to participate in exciting activities such as skydiving, whereas people who want to feel low-arousal positive affect would like to participate in relaxing activities such as hot spring bathing. In Tsai's (2007) research on ideal affect, she found that European American undergraduate students preferred high-arousal positive affect and participated in active sports more often compared to Asian American students who preferred low-arousal positive affect. Thus, Tsai integrated ideal affect, or what one wants to experience emotionally, with the types of activities one wants to engage in and demonstrated that this association may be culturally influenced. Recently, Tsai's proposition was supported by Mannell *et al.* (2014) who reported that during leisure, but not paid work, Euro-Canadian adult males realised their ideal levels of low-arousal positive affect.

Given the close relationship between leisure and sport tourism, affect valuation theory also helps to understand the motivation of sport tourists. Tsai (2007) reported that when European American undergraduate students were asked about their ideal vacations, they described more active sports participation (e.g. surfing, running) than Hong Kong Chinese undergraduate students. Another example is found in the context of outdoor recreation activities that some people pursue during their leisure travels. According to Walker *et al.* (2001), European North-American outdoor recreationists preferred high-arousal activities (e.g. hiking), whereas Chinese outdoor recreationists preferred low-arousal activities (e.g. viewing scenery). Therefore, it is reasonable to suspect that our ideal affect would be key to examine sport tourists' motivation generally but more particularly in cross-cultural contexts.

Global international tourist arrivals reached 1186 million in 2015 and are expected to reach 1.8 billion by 2030 (World Tourism Organisation, 2016). Asian countries are following the same trend as other tourism destinations. Japan is becoming an increasingly popular tourist destination with the number of international inbound tourists increasing beyond 24 million (Japan National Tourism Organisation, 2017). Although most inbound tourists come from China, South Korea and Taiwan, among European and North American countries, the United States became the first country to exceed 1 million tourists to

Japan in 2015 and reached over 1.2 million in 2016 (Japan National Tourism Organisation, 2017). This pattern indicates that the cultural backgrounds of international inbound tourists in Japan have become diverse. The current situation of inbound tourism in Japan warrants a close examination of the relationships that link tourism, motivation and culture, which will provide important practical implications for the Japanese sport tourism industry.

Affect valuation theory implies that one should consider the cultural backgrounds of tourists to ensure effective tourism marketing. Of course, this applies for sport tourism in Japan as well as other types of tourists. First, marine sport activities are an important form of active sport tourism, given Japan's abundant marine resources which include over 6000 islands (Harada, 2016). As Tsai (2007) reported, Westerners would be expected to be interested in physically rigorous marine activities like surfing, whereas East Asians would be more likely to be interested in relaxing marine activities like snorkelling. Given that surfing will be introduced as a demonstration sport in the Tokyo 2020 Olympics, Japan will likely capture Westerners' attention as a surf tourism destination.

Second, Japanese baseball games provide an interesting insight into contrasting spectator event sport tourism motivations and experiences. Westerners would, according to Tsai's (2007) theorisation, be interested in watching baseball games from outfield seats with private cheering groups (i.e. unique Japanese style of collective cheering with trumpets or drums), whereas East Asians would be interested in watching the game from luxury infield seats far away from the cheering groups. In addition to the exciting cheering groups, (a) launching balloons at seventh-inning and (b) buying beer from 'beer-girls' who wear highly sexualised uniforms and roam the stands with beer tanks on their backs while smiling and waving to potential customers would also provide high-arousal experiential states to Western tourists at Japanese ballparks (Graczyk, 2014).

Lastly, with regard to heritage sport tourism, Westerners are likely to be particularly motivated to visit interactive museums with cutting-edge interpretive techniques, designs and technologies, whereas East Asians tend to be more interested in traditional museum-style presentations surrounded by a calm atmosphere. For example, a world-known Japanese sporting company, ASICS, owns ASICS Sports Museum in Kobe, which is regarded as a popular tourist destination. Given that this museum offers an interactive and fully immersive experience of virtual athletics through new technologies (ASICS, n.d.), this sport heritage facility would be more appealing to Westerners than East Asians. While theoretically informed, no scientific research has been directed towards

(Continued)

Case Study 3.1: (Continued)

empirical investigation of these possibilities; therefore, more research is clearly warranted.

In conclusion, Tsai (2007: 254) stated that 'Ideal affect … directs people's behavioral choices'; if so, a fuller understanding of sport tourist motivation can only be achieved if attention is paid to cultural contexts. If we can predict what tourists want to feel when they partake in leisure travel based on their cultural background, we would be able to develop more efficient and effective marketing strategies generally and particularly for the sport tourism industry. This would open the way for the creation of targeted sport event tour packages, inform event design for active sport participants and allow for market-specific targeted marketing campaigns to support events such as the Japan 2019 Rugby World Cup, the 2020 Tokyo Olympic Games and the 2021 Kansai World Masters Games.

Key Reading

Ito, E. and Walker, G.J. (2016) Cultural commonality and specificity in Japanese and Euro-Canadian undergraduate students' leisure experiences: An exploratory study on control and positive affect. *Leisure Sciences* 38, 249–267.

Tsai, J.L., Chim, L. and Sims, T. (2015) Consumer behavior, culture, and emotion. In S. Ng and A.Y. Lee (eds) *Handbook of Culture and Consumer Behavior* (pp. 68–98). New York: Oxford University Press.
   Additional references cited in this case study are included in the Reference list.

Behaviouristic market segmentation

This avenue of segmentation classifies consumers according to their behavioural relationship with a product (Swarbrooke & Horner, 1999), with implications for the visitor experience (Chapter 8). Millington *et al.* (2001), for example, profile the growing number of participants in adventure tourism dividing the market based on soft (e.g. cycling, canoeing and horse riding) and hard (e.g. rafting, kayaking, climbing and caving) adventure activities. These adventure sports can be further differentiated based on the behaviours of participants. Downhill mountain bike racing and white-water kayaking are extreme versions of sport that can be segmented based on the motivations and behaviours of participants, and then profiled demographically (Millington *et al.*, 2001). The link between motivation and behaviour, which is well established in the sport and recreation literature (Jackson, 1989), is of high relevance to sport tourism. It is important for destination marketers to understanding the motivational and behavioural profiles of sport tourism market segments. They determine the desired visitor experience and the secondary activity sets that are associated with members of specific tourism market segments (Nogawa *et al.*, 1996).

The sport of skiing provides an illustration of the varied motivations and behaviours that exist within the tourist market (Klenosky *et al.*, 1993). Richards (1996), for instance, analysed the extent to which British skiers are motivated by technical challenge and the enhancement of their skiing ability. This research identified a market segment that was motivated by challenging skiing experiences. The quality of the ski conditions and varied terrain were found to be fundamental to the experiences sought by this market segment (Richards, 1996). By contrast, the decision-making process of less experienced skiers was influenced more by price and accommodation. Participants in scuba diving (Tabata, 1992) and sport fishing (Roehl *et al.*, 1993) give increasing priority to the quality of the sport experience over other aspects of the visitor experience as they become more experienced in their sports. These studies confirm the importance of motivations and behavioural profiles when undertaking sport tourism market analyses (Davies & Williment, 2010).

Heritage Sport Tourism

Within our fourfold classification, heritage sport tourism is the least researched, although some important contributions to the published literature now exist (Gammon & Ramshaw, 2013). This form of sport tourism includes tourist visitation to sport museums, halls of fame, themed bars and restaurants, heritage events and sports reunions (also see Chapter 10). Heritage is a rapidly developing aspect of sport tourism. Gammon (2002) discusses the commercialisation of the past in tourism and relates this to sport tourism. He documents the growth of heritage sport tourism with reference to the mature sport nostalgia industry in North America (Gibson, 2002; Redmond, 1990). The resource base for heritage sport tourism focuses particularly on halls of fame and sport museums. While the former venerate the famous, the gifted or the exceptional, the latter contain collections of artefacts and memorabilia that celebrate a sport, rather than high performing individuals or teams within a sport (Gammon, 2002). The growth in demand for sport heritage extends to stadium tours and various other forms of heritage engagement (Gammon & Fear, 2005; Ramshaw & Gammon, 2010), although this growth in tourist demand is not yet reflected in a large published body of research on this topic.

Nostalgia represents an avenue of sport tourism that has strong parallels with heritage tourism (Redmond, 1990). Bale (1989) notes that sport edifices may develop such a mystique that they become subjects of visitor attention in their own right. Examples of sports venues that have powerful heritage values include Wembley Stadium and the Wimbledon

Lawn Tennis Club (London), the Athens (1896) and Berlin (1936) Olympic stadiums and the Holmenkollen ski jump (Oslo) (Bale, 1989). Numerous questions relating to sport heritage remain unanswered. The need exists to understand why people engage in this form of tourism and how nostalgia relates to other forms of heritage tourism and sport tourism at a destination.

The focus of research in this field has hitherto examined sport heritage resources (Redmond, 1990) rather than how and why people engage in this form of sport tourism (Gammon, 2002), although this has started to change (Lamont & McKay, 2012; Morgan, 2007; Ramshaw & Gammon, 2010). How sport heritage relates to demand for active and event sport tourism experience is also poorly understood. It has become increasingly clear that this form of sport tourism has indistinct boundaries with spectator events, participation events and active engagements in recreational sports. Travel packages that follow the tour matches of an international sports team, and which are often led by former star players, illustrate the overlap between spectator event and sport heritage. Similarly, cruise ship packages that offer passengers the opportunity to meet or be coached by sports personalities (Gibson, 1998a), hold elements of both active engagement and nostalgia. Recent research on cycling events has shed light on the complex interplay of elements of event attendance, active participation and embodiment, and nostalgia (Derom & Ramshaw, 2016; Lamont & McKay, 2012) (see Focus Point 3.2). An increasingly apparent trend over the last decade has, in fact, been the development of sports events that embrace the widest possible spectrum of niche markets, incorporating elements of competition (elite and non-elite), active participation, spectatorship and heritage. Focus Point 3.2 points towards a blurring of different forms of engagement in sports, and multimodal experiences in many forms of contemporary sport tourism.

Focus Point 3.2: Postmodern Sports Tourism at the Tour de France

Matthew Lamont

Organised tours to the Tour de France (*Le Tour*) often augment spectating opportunities with active cycling excursions. These emerging, multimodal experiences highlight a need for renewed theoretical perspectives to encapsulate evolving sports tourism experiences. Applied to tourism, the perspective of postmodernism challenges modernist theorisations of tourism experiences as homogeneous with respect to tourists' motivations, experiences and environments (Uriely, 1997). Postmodernism emphasises contemporary tourists' quests for authenticity pursued through multifaceted touristic behaviours imbued with mobility, embodiment and nostalgia. Moreover, validation of media-shaped constructions of place attained through co-presence with sites of personal, cultural or historical significance may yield profound, intensely fulfilling experiences.

Qualitative data were collected from participants on a cycling and spectating tour of *Le Tour* 2011 to explore their motives and consumptive practices. Kinaesthetic, mobile engagement with *Le Tour*'s landscapes permeated narratives around their motives for selecting this tour. Most spoke of having followed *Le Tour* on TV for many years and conveyed palpable excitement around cycling its landscapes. Constructed around the 2011 event's crescendo stages in the French Alps, a mobile schema enabled participants to cycle parts of *Le Tour*'s route featuring iconic mountain passes such as Alpe d'Huez and the Col du Galibier, 'divine and pitiless backdrops' (Gaboriau, 2003: 69) where professional cyclists have etched their name into *Le Tour* folklore. Participants then fluidly segued into spectator roles, watching the race entourage pass by before being transported by coach back to the hotel at day's end.

While the French Alps were visually striking, participants desired more than merely gazing at imposing landscapes. Intimations of postmodern tourism were evident in that these cycling fans yearned authentic, self-actualising and physically stimulating experiences. Kinaesthetic submergence in *Le Tour*'s landscapes was a common motive and was crucial to reconciling *a priori* constructions of *Le Tour*'s reality. As one participant described, 'There are certain great ones [mountains] which have so much history of significance in previous Tours … it would be just like a pilgrimage for me to be able to say I've ridden where X attacked Y'. Happenstance encounters with *Le Tour* cyclists further enhanced the authenticity of their experiences as did co-constructing roadside communitas among large crowds of spectators. Participants' performative acts thereby created 'place', endowing those sites with 'hot' authenticity (Cohen & Cohen, 2012; Lamont, 2014).

This study highlights limitations in seminal sports tourism typologies that downplay potentially fluid, multifaceted roles that may be adopted by contemporary sports tourists. A postmodernist lens is beneficial in encapsulating multifarious motivations driving sports tourism participation while simultaneously accounting for diverse modes of consumption imbued with mobility and embodiment in the pursuit of fulfilling, authentic experiences.

Key Reading

Lamont, M. and McKay, J. (2012) Intimations of postmodernity in sports tourism at the Tour de France. *Journal of Sport & Tourism* 17 (4), 313–331.

Additional references cited in this focus point are included in the reference list.

Conclusion

This chapter confirms that sport tourism may be accurately conceptualised as a specialised market in itself while also being characterised by a diversity of niche markets (Bull & Weed, 1999; Chalip, 2001; Higham & Hinch, 2009). These niche markets can be differentiated through geographic, socio-economic, demographic, psychographic and behaviouristic segmentation techniques, and vary on cultural grounds. The relevance of critical sport tourism market insights lies in the diversity of the constituent segments that collectively comprise the sport tourism market. The travel profiles (e.g. length of stay, modes of transport, accommodation preferences) and secondary tourism motivations (e.g. attractions and activities) that these travellers bring to the tourism destination are the subject of tourism market research. Scholarly research that addresses sport and tourism as it relates to different niche markets is critically important to our understanding of sport tourism phenomena, and the effective and sustainable development of sport-related tourism at a destination (Hinch *et al.*, 2016).

The study of sport tourism markets offers rich and timely avenues of academic research. The travel motivations and preferences of distinct sport tourism markets, and the tourism development opportunities that they offer, have become subjects of a developing academic literature. Glyptis (1991: 181) was one of the first to observe that sport tourism development necessitates that 'sport and tourism authorities talk to one another and forge real working partnerships'. Detailed insights into sport tourism markets, that extend across spectator and participation events, to active engagement in recreational sports and sports heritage, will help to facilitate the establishment of such partnerships.

4 Development Processes and Issues

Management of the process of renewal and the redesigning of products and services is a field where sport and tourism can exchange valuable experiences.
de Villers, 2001: 12

Introduction

Change is one of the few constants in contemporary society. By understanding the nature and processes of development, trends can be identified and potentially influenced. This chapter discusses the concept of development in relation to sport tourism. It is argued that if sport tourism development is to comply with the paradigm of sustainability, then planned and informed intervention into the development process is required. The latter part of this chapter highlights three key issues facing sport tourism development including commodification/authenticity, globalisation and organisational fragmentation. Failure to understand these issues compromises the development potential of sport tourism. Brendon Knott's case study of sport tourism development in South Africa illustrates many of the opportunities and challenges that sport tourism presents as an agent for development (Case Study 4.1).

The Concept of Development

'Development' is an elusive term with different meanings in different contexts. Members of a broad range of disciplines and professions interpret development in ways that make sense to themselves, perhaps more so than others. Even in the context of popular usage, fundamentally different meanings of the term exist. Common interpretations include development as philosophy, process, plan and product (Sharpley, 2014). As a philosophy, development alludes to a desired future state. As a process, it refers to the way a society moves from one condition to another. This particular condition or state reflects the product view of development while the articulation of actions to reach desired states is a manifestation of development as a plan. While these two perspectives are noteworthy, the most relevant perspectives of development for this chapter are development as a product and development as a process. In the case of the former, development is treated as a state. This approach is being used when reference is made to 'levels of development'. These levels were traditionally assessed in terms of economic measures such as income and employment but increasingly include non-economic measures such as social conditions.

Planners and policymakers seek change that is accompanied by positive impacts and implications. Growth in its various forms, including the number of sport-related visitors to a destination, the amount that these visitors spend and the physical development of sport tourism sites, are often key objectives. Nevertheless, it is wrong to simply equate growth with development, especially when a long-term perspective is taken that incorporates multiple stakeholder interests (Atkisson, 2000). Development is not just change in terms of growth, it is change that has a positive overall impact. It should, therefore, be assessed in the context of a full range of stakeholders and their often competing and sometimes contradictory goals.

While there is agreement that development as a planned process should be aimed at positive change over time, there has been less agreement as to what should be used to measure such change. Typically, economic measures have been used as indicators. This is consistent with the dominant political-economic system operating throughout the developed world, which focuses on economic growth as an unassailable policy goal (Hall, 2008). Notwithstanding this entrenched perspective, definitions of development 'have tended to be broadened over time and development has gradually come to be viewed as a social as well as an economic process which involves the progressive improvement of conditions and the fulfilment of potential' (Wall, 1997: 34). For example, poverty reduction has increasingly been recognised by UN agencies as a key development objective (Hawkins & Mann, 2007; Rogerson, 2014). Clearly, there has been a shift over time in the way development has been understood, from a narrow focus on economic measures to a broader focus that, according to Sharpley (2014), includes:

(1) economic: wealth creation and equitable distribution;
(2) social: progress in health, education, employment and housing;
(3) political: human rights and political freedom;
(4) culture: protection of identity and self-esteem;
(5) full-life paradigm: strengthening of meaning systems, symbols and beliefs;
(6) ecology: environmental sustainability.

For the purpose of this book, we have adopted Sharpley's (2014: 22) definition of development as '...the continuous and positive change in the economic, social, political and cultural dimensions of the human condition, guided by the principle of freedom of choice and limited by the capacity of the environment to sustain such change'.

Sustainable Sport Tourism Development

Sustainable development is a contested concept. While few people would claim that they do not support the principle of sustainable development, the meaning of the concept is not universally shared (Hopwood *et al.*, 2005; Pesqueux, 2009). Business advocates tend to emphasise 'development' while environmentalists tend to focus on 'sustainability'. As a result, the phrase can sometimes seem like a contradiction of terms. Yet, behind the rhetoric of the various stakeholders and scholars who debate its merits, it is a concept that holds widespread appeal.

The World Commission on Environment and Development's (WCED) (1987: 4) definition of sustainable development is '...development that meets the needs of the present without compromising the ability of future generations to meet their own needs'. While there are many other definitions, most early definitions tended to focus on the sustainability of natural ecosystems. The WCED definition is, however, robust enough to include the sustainability of other types of resources including culture, which Mowforth and Munt (2015) describe in terms of tourism hosts being able to maintain their cultural identity. In the case of sport tourism, this aspect of the definition encompasses sport as an expression of culture (Chapter 6) as well as the impact that sport tourism has on the physical environment in which it occurs (Chapter 7).

There is a difference between sustainable tourism and sustainable development (Liu, 2003; Saarinen, 2006; Zhu, 2009). Sustainable tourism is '...tourism which is in a form which can maintain its viability in an area for an indefinite period of time' (Butler, 1993: 29). In contrast, tourism in the context of sustainable development is tourism that is developed and maintained in an area (community, environment) in such a manner and at such a scale that it remains viable over an indefinite period and does not degrade or alter the environment (human and physical) in which it exists to such a degree that it prohibits the successful development and well-being of other activities and processes (Butler, 1993). In their editorial summing up the first 25 years of the *Journal of Sustainable Tourism*, Bramwell *et al.* (2017: 1) report that the concept of sustainable tourism has evolved to where it is 'now seen as a normative orientation that seeks to re-direct societal systems and behaviour on a broad and integrated path toward sustainable development'. In the process of this evolution, it has come to embrace social, cultural, economic and political dimensions alongside environmental ones. Accounting for this broader development context is challenging but necessary. The achievement of sustainable sport tourism requires a balance and, at times, trade-offs between goals in these various domains of sustainability.

While sustainable sport tourism is the essence of Figure 4.1, the broader concept of sport tourism in the context of sustainable

SOCIAL/CULTURAL GOALS   ECONOMIC/POLITICAL GOALS

ENVIRONMENTAL GOALS

Figure 4.1 Sustainable sport tourism (Source: After Hall, 2007)

development is embodied in the spheres adjoining the centre. A healthy sport tourism economic/political sphere should ideally support and enhance the social/cultural dimension of the community. It should also play a similar role in the context of the natural environment, which features prominently in many types of sport tourism activity. In addition, sport tourism social/cultural practices should serve as positive forces in relation to the natural environment. There is, however, no guarantee that the interaction between sport and tourism will necessarily be positive. To achieve positive outcomes, those with management and planning interests in sport tourism must be conscious of the impacts of their decisions throughout the full range of these realms rather than just at the centre. This awareness needs to be accompanied by a constructive integrated approach to development.

The sport and tourism industries have vested interests as well as a moral obligation to meet the goal of sustainability. At a micro level, sustainable development has a direct impact on the return on investment for sport tourism businesses and the communities in which they function.

At a macro level, sustainable development has global implications across an intricate web of social, economic and environmental domains (Focus Point 4.1). The natural tendency of developers is to focus on economic goals, but if environmental and social cultural resources are viewed as a form of capital, a strong business argument exists for sustainable practices (Hall, 2008). Beyond the business rationale, a moral responsibility exists for the pursuit of sustainable development at both the micro and macro levels (Pawłowski, 2008). The global relevance of sustainable tourism was highlighted by the United Nations designating 2017 as the International Year of Sustainable Tourism for Development. This designation, along with the UN's Sustainable Development Goals (Table 4.1), provides an ambitious target for sectors like sport tourism.

Focus Point 4.1: Sustaining Participatory Events – The Case of the Canadian Death Race (CDR)

The CDR is an ultramarathon event held in the small isolated community of Grande Cache (GC), Canada, which has a population of 4000 (Hinch, 2013). The event is appreciated in GC for putting the community 'on the map' and providing a tourism initiative that helps to offset the cyclical nature of its resource extraction-based economy. The race features a course of five stages totalling 125 km in distance, three mountain summits and 5181.6 m in elevation change, all of which must be completed within 24 hours (CDR, 2017). From its humble beginnings in 2000, participation in the race peaked at just over 1600 registered runners in 2012. These runners and their accompanying support crews essentially double the population of GC on race weekend. Since 2012, registrations have decreased marginally due to increased completion of other events and the challenges of burnout and fatigue within GC. Benefits have accrued in all major realms of the community including: (1) the injection of substantial revenues for the accommodation and food & beverage industries; (2) increased community pride as the community has reaped accolades for hosting such a successful event; and (3) a greater appreciation of the beauty of the surrounding landscape and natural setting. Despite these benefits, the sustainability of the event is being challenged on a number of fronts. This is particularly true in the social/cultural realm. While the flood of visitors on race weekend is appreciated by many residents for the energy it brings, others resent the disruption to their normal routine. This reaction is not surprising given the magnitude of the influx of visitors. A more pressing challenge has been that of volunteer burnout. While community groups and individual residents were keen to help out at the inception of the event, their participation has steadily dropped off

(Continued)

Focus Point 4.1: (Continued)

as the demands have increased with the growth of the event, volunteers have simply tired of ongoing participation and concerns have been raised about the equity of the distribution of the event revenues. This challenge has been partially addressed by the municipality requiring its employees to help out with the event as part of their work duties and by the recruitment of volunteers from outside of the community (a type of sport tourist that is poorly understood). The sustainability of the CDR and other participant-based events of this scale in similar sized communities will be closely tied to whether such strategies prove to be successful in the long run.

Key Reading

CDR (online) (2017) The Canadian Death Race. See http://www.canadiandeathrace.com/home.html (accessed 29 August 2017).

Hinch, T.D. (2013) Ultra-marathons and tourism development: The case of the Canadian Death Race in Grande Cache, Alberta. In B. Garrod and A. Fyall (eds) *Contemporary Cases in Sport* (pp. 22–40). Oxford: Goodfellow.

Two special issues of the *Journal of Sport & Tourism* have made important contributions to our understanding of sustainability in a sport tourism context. The first was Fyall and Jago's (2009) special issue on sport tourism sustainability. Key insights offered in this special issue included: (1) recognition of the need to move from rhetoric to substance; (2) the increasingly widespread adoption of a triple bottom line approach to sustainability; and (3) recognition of the fact that sport tourism not only produced impacts that affected sustainability in destinations but that it was also a recipient of impacts from external sources which had a direct influence on sport tourism activities. The second special issue of relevance was Hinch *et al.*'s (2016) special issue on sport tourism destinations in the context of sustainability. Contributions to this special issue examined the sustainability of sport tourism destinations: (1) at differing scales of activity and destination (Derom & Ramshaw, 2016); (2) single versus multisport destinations (e.g. Carneiro *et al.*, 2016); (3) publically managed resources (e.g. Halpenny *et al.*, 2016); (4) emerging sports (Moularde & Weaver, 2016) versus traditional sports (Evans & Norcliff, 2016); (5) seasonality (Hodeck & Hovermann, 2016); and (6) theory with examples including Usher and Gomez's (2016) contributions on 'territoriality' in the case of surfing and Bunning and

Table 4.1 UN Sustainable Development Goals (2015–2030)

(1)	No poverty	End poverty in all its forms everywhere
(2)	Zero hunger	End hunger, achieve food security and improved nutrition and promote sustainable agriculture
(3)	Good health and well-being	Ensure healthy lives and promote well-being for all at all ages
(4)	Quality education	Ensure inclusive and quality education for all and promote lifelong learning
(5)	Gender equity	Achieve gender equality and empower all women and girls
(6)	Clean water and sanitation	Ensure access to water and sanitation for all
(7)	Affordable and clean energy	Ensure access to affordable, reliable, sustainable and modern energy for all
(8)	Decent work and economic growth	Promote inclusive and sustainable economic growth, employment and decent work for all
(9)	Industry, innovation and infrastructure	Build resilient infrastructure, promote sustainable industrialisation and foster innovation
(10)	Reduced inequalities	Reduce inequality within and among countries
(11)	Sustainable cities and communities	Make cities inclusive, safe, resilient and sustainable
(12)	Responsible consumption and production	Ensure sustainable consumption and production patterns
(13)	Climate action	Take urgent action to combat climate change and its impacts
(14)	Life below water	Conserve and sustainably use the oceans, seas and marine resources
(15)	Life on land	Sustainably manage forests, combat desertification, halt and reverse land degradation, halt biodiversity loss
(16)	Peace, justice and strong institutions	Promote just, peaceful and inclusive societies
(17)	Partnerships for the goals	Revitalise the global partnership for sustainable development

Source: UN. Sustainable Development Goals. See http://www.un.org/sustainabledevelopment/ sustainable-development-goals/ (accessed 29 August 2017).

Gibson's (2016) examination of the role of travel conditions in the context of active sport tourism events.

Planning

Planning is a means of managing change. Given that sport tourism exists in a dynamic environment and given that these processes do not necessarily lead to sustainable outcomes, some sort of intervention in the development process is required to foster sustainability. Essentially, 'planning is a process of human thought and action based upon that thought…' (Chadwick, 1971: 24). Planning is 'concerned with anticipating and regulating change in a system, to promote orderly development as to increase the social, economic, and environmental benefits of the development process' (Murphy, 1985: 156). It is based on the assumption that even a partial understanding of the dynamics of sport tourism and the world in which it exists, provides the basis to influence change. By consciously initiating this type of process, developers of sport tourism not only act in their best self-interests, but they make positive contributions to the sustainability of the social, cultural, economic and environmental systems in which they function.

The underlying process for planning is consistent across a broad range of fields and disciplines. It is based on an assessment of the current situation, likely changes that will occur in the environment in which the plan is being conducted, decisions on the desired end state, formulation of some sort of action plan, its implementation, followed by monitoring, assessment and adjustment as required (Esfahani *et al.*, 2009). Bagheri and Hjorth (2007) argue that 'process-oriented' planning is more important than 'goal-oriented' planning because sustainable development is an ideal not an absolute end point. From this perspective, the true benefit lies in the social learning process of engaged stakeholders. Yang and Wall (2009) illustrate this with their study of sustainable planning for ethnic tourism in Xishuangbanna, China. They make the point that planners must deal with dynamic issues, which require ongoing resolution rather than one-time attention. While Yang and Wall drew their conclusions in the context of ethnic tourism, parallel issues that require continuous attention exist in the realm of sport tourism.

Development Issues

If planning interventions in the development process are to be successful, then informed consideration of the many issues that exist within the field of sport tourism is necessary. A number of specific issues are introduced and discussed in the chapters that follow, but three issues have particular significance to sport tourism development. These issues include commodification/authenticity/authentication, globalisation and organisational fragmentation.

Commodification, authenticity and authentication

One of the most fundamental issues of sport tourism development concerns the process of commodification and its implications in terms of authenticity. This issue is predicated on the belief that the search for authenticity is one of the main driving forces for tourists (MacCannell, 1976; Urry, 1990) and that sport represents a dynamic and increasingly prominent tourist attraction (Hinch & Higham, 2001).

Cohen (1988) defined commoditisation (commodification) as:

a process by which things (and activities) come to be evaluated primarily in terms of their exchange value, in a context of trade, thereby becoming goods (and services); developed exchange systems in which the exchange value of things (and activities) is stated in terms of prices form a market. (Cohen, 1988: 380)

Similarly, in the context of sport, McKay and Kirk (1992) defined commodification as:

the process by which objects and people become organized as things to be exchanged in a market. Whereas cultural activities such as ... sport once were based primarily on intrinsic worth, they are now increasingly constituted by market values. (McKay & Kirk, 1992: 10)

These definitions feature the same fundamental characteristics: a process of commercialisation that superimposes economic values on things or activities that were not previously valued in this way.

Tourism critics argue that the commercialisation of culture introduces economic relations into an area where they previously played no part. In the process of commercialisation, real authenticity is destroyed and a covert 'staged authenticity' emerges. As the fact that this staged authenticity is not real dawns on tourists, it thwarts their genuine search for authenticity (MacCannell, 1973; Yang & Wall, 2009). Cohen (1988), however, argued that commoditisation does not necessarily destroy the meaning of a cultural attraction although it may change it.

Critics of the commodification of sport suggest that the professionalisation of various competitions and the broader commercialisation of sport through the media and the interests of large manufacturing/retail corporations have had a detrimental effect. Stewart (1987) articulates this position by arguing:

social hegemony of the commodity form is apparent as the practice of sport is shaped and dominated by the values and instrumentalities of a market. ...the idealized model of sport, along with its traditional ritualized meanings, metaphysical aura, and skill democracy, is destroyed as sport becomes just another item to be trafficked as a commodity. (Stewart, 1987: 172)

A particularly good illustration of this line of argument is the commodification of the Olympic Games. Starting as a strictly amateur event that was envisioned as an instrument for international goodwill, the Games have evolved into a global commercial giant of imposing power and influence (Barney et al., 2002). In a similar vein, Morgan (2014) argues that the commodification of place in tourism has often undermined culture and identity in popular tourist destinations. Clearly, the commodification of sport and places is not always problematic as it also brings benefits in the economic, social and environmental realms.

Authenticity is closely related to commodification. Cole (2007) distinguishes between the perspectives of government, tourist and villagers in Eastern Indonesia. While governments and tourists found authenticity to be problematic, villagers saw the commodification of their culture as empowering. In a tourism context, authenticity has traditionally been viewed in relation to the object of interest, the originals or the thing or activity that the tourists have come to see (Wang, 1999). This can be thought of as comparable to a museum curator's perspective where experts in such matters test whether objects of art are what they appear to be or are claimed to be (Cohen, 1988). The sporting parallel is manifest in popular criticism of any break from tradition, especially in relation to rule changes. From this perspective, a thing or activity is judged to be authentic or not in a relatively objective manner. Reisinger and Steiner (2005) take issue with this view, arguing that it is not possible to judge object authenticity. In the face of such criticism, the pursuit of objective authenticity has gradually given away to more flexible interpretations of authenticity where it is appreciated that there is seldom an absolute authenticity, but that it is more often negotiated in some fashion. Wang (1999) has termed this as 'constructive authenticity'. As part of the commodification process, the tourism industry has continued to present 'staged authenticity', albeit increasingly in an overt form. Rather than taking tourists to the backstage of a destination to experience real culture, tourism operators use the front stage where a destination's culture is presented in a controlled way via museums, heritage centres, cultural performances and other similar forums. A dedicated stage for tourism allows for greater influence over impacts and addresses the operational constraints associated with tourist activity such as the restricted scheduling that typically characterises tourists.

Timothy and Boyd's (2002) review of the authenticity debate in the context of heritage tourism highlights several important issues that are especially relevant in relation to heritage sport tourism. They suggest five types of common distortions of the past including: invented places or reconstructed pasts; relative authenticity that recognises the subjective nature of interpreting the past; ethnic intruders in the form of non-local interpreters; sanitised and idealised pasts; and finally, the unknown past in which it is recognised that interpretations of the past can only be partial. Sport tourism examples in these categories include: fantasy sport camps, contradictory views of historic sporting matches, great geographic diversity reflected in the playing and coaching rosters, highly nostalgic interpretations of past sporting glory and hero worship, and memories of the past built on selected statistical summaries associated with elite competition (Gammon & Ramshaw, 2007). All of these distortions compromise objective authenticity but they do not necessarily detract from the experience of the visitor.

While many academics have criticised the tourism industry's failure to provide 'objective authenticity' (e.g. Boorstin, 1975; Greenwood, 1989), others recognise that tourists often seek contrived experiences as part of their desire to have fun (Moscardo, 2000; Urry, 1990). Notwithstanding this perspective, it is acknowledged that many tourists are searching for authentic experiences rather than authentic objects (McIntosh & Prentice, 1999; Tolkach *et al.*, 2016). This is consistent with the more general consumer search for authenticity in products, services and experiences (Gilmore & Pine, 2007; Yeoman *et al.*, 2007). Wang (1999) has articulated this 'rethinking' of authenticity, describing an activity-related authenticity which he refers to as existential authenticity:

> Existential authenticity refers to a potential existential state of Being that is to be activated by tourist activities. Correspondingly, authentic experiences in tourism are to achieve this activated existential state of Being within the liminal process of tourism. Existential authenticity can have nothing to do with the authenticity of toured objects. (Wang, 1999: 352)

This type of authenticity would seem to hold considerable relevance for sport tourism with its focus on experience. It also represents an interesting dynamic in relation to the blurring of sport and entertainment. The key is engagement. If tourists are actively engaged as spectators or as active participants in a sport, they are likely to view their experience as being authentic regardless of how others may assess the situation. Sport is unique compared to other types of tourist attractions in this regard. Key characteristics of sport that promote experiential authenticity include: uncertain outcomes, display as part of performance, its physical basis and all-sensory nature, self-making and the construction of identity, and its propensity to develop community. To a large extent, authenticity is what makes sport so compelling (Hinch & Higham, 2005).

Wang (1999) expands on his interpretation of existential authenticity by introducing four subcategories. The first of these is intra-personal authenticity as manifest in bodily feelings – a major dimension of participatory events and active sport tourism. Intra-personal authenticity as manifest in self-identity is the second subcategory. This category is described as 'self-making' and is often found in extreme and lifestyle sports in which participants can reaffirm and develop their sense of identity. A third type is inter-personal authenticity that enhances family ties. In a sporting context, family ties are closely related to the concept of a team. Finally, inter-personal authenticity can be found in the vibrant communities associated with various sports including fandoms and sport subcultures.

Despite an intriguing and active academic discourse on tourism authenticity, a consensus has not emerged. In reflecting on this situation, Cohen and Cohen (2012: 1296) reframed the focus to one of authentication or the social processes through which the authenticity of the attraction is formed. They defined authentication as 'a process by which something – a role, product, site, object or event – is confirmed as "original", "genuine", "real" or "trustworthy"'. They further characterised this concept into 'cool' and 'hot' authentication with cool authentication being closely related to Wang's (1999) concept of objective. Fundamentally, cool authentication is guided by experts who declare an object, site, event, custom, role or person to be authentic based on scientific knowledge. Such declarations are often contested but are commonly employed in tourism contexts. In contrast, hot authentication is more closely aligned with Wang's (1999) concept of experiential authenticity. Cohen and Cohen (2012: 1301) describe hot authentication as 'an affective self-reinforcing process in which the sacredness, sublimity, or genuineness of sites, objects or events is constantly perpetuated, confirmed (and augmented) by public practice, rather than by some declaration'.

Lamont's (2014) study on authentication in sport tourism was one of the first empirical studies to examine this process. In his study of the French Alps as a significant landscape in the Tour de France, Lamont used the context of a commercial tour that included opportunities to cycle parts of the route just prior to the professional competitors to study the authentication process. Three key categories of authentication emerged from his participant observations and interviews. The first was authentication through 'embodied athletic performances' as experienced through the act of performing a similar activity in the same place that professional riders in the Tour de France perform. In this case, authentication went beyond the tourism gaze that reflects the tourism industry's promotional messages (Urry, 1990) to one of a full range of sensory experiences as described by Spinney (2006) in a similar cycling experience. The second authentication process involved 'collective roadside practices' during which the sport tourists became part of the festive atmosphere of the Tour. A state of communitas or celebration at the level of both the tour group and the race itself broke down the normal social barriers and fostered an authentic experience. The third authentication process that Lamont (2014) found was the 'mediation of touristic encounters with place'. In particular, the ways that these sport tourists communicated their experiences on their tour through photographs and social media proved to be a powerful way for them to authenticate their experience on the mountain ranges of the Tour de France.

Globalisation

The term 'globalisation' can bring forth a variety of reactions as exemplified by the positive response to the concept of a 'global village' through to the passionate protests decrying economic inequities and cultural imperialism that have become a predictable part of world economic summits. Simply put, globalisation is the process that leads to an ever-tightening network of connections, which cut across national boundaries. It is characterised by the worldwide compression of time and space (Mowforth & Munt, 2015). Sport is increasingly being used as a way to position destinations within this global world (see Case Study 4.1).

In many ways, globalisation is a new way of looking at development. Globalisation is manifest in a web of political, economic, cultural and social interconnections (Go, 2004; Lew, 2014; Milne & Ateljevic, 2004). It is no longer possible for communities to function in isolation from other parts of the globe. Globalisation is a complex process and a variety of interpretations and perspectives have emerged (Silk & Jackson, 2000). The cultural imperialist interpretation sees local culture as being displaced by a foreign one, causing a homogenising trend and the creation of a common global culture. This form of imperialism is often equated with Americanisation, as the US economy and culture are viewed as the dominant forces in the global system although they are currently being challenged by China's growing economy. A cultural hegemony interpretation sees globalisation as a two-way process. Local communities receive global images, goods and services but interpret them on their own terms. In this two-way relationship, local groups play a key role in the way global trends are manifest in their locale (Whitson & Macintosh, 1996). Finally, the figurational perspective advocates a view of globalisation as a process. This perspective emphasises a long-term historical approach that involves the examination of cases of domination and resistance over time. Maguire's (1999: 3) definition of globalisation fits this interpretation.

Globalisation processes are viewed here as long-term processes that have occurred unevenly across all areas of the planet. These processes – involving an increasing intensification of global interconnectedness – appear to be gathering momentum and despite their 'unevenness', it is more difficult to understand local or national experiences without reference to these global flows. Every aspect of social reality – people's living conditions, beliefs, knowledge and actions – is intertwined with unfolding globalisation processes. These processes include the emergence of a global economy, a transnational cosmopolitan culture and a range of international social movements.

Case Study 4.1: Leveraging Sport Tourism Events as Catalysts for Sustainable Place Brand Development – South Africa and the 2010 FIFA World Cup

Brendon Knott, Cape Peninsula University of Technology

Sport tourism events have been frequently coupled with urban regeneration and event-linked or event-themed sustainable development projects. Beyond tourism benefits such as improving accommodation capacity and service delivery, these investments are often aimed at signalling or reinforcing a nation's event hosting and project management capability and enhancing its global reputation. This has been particularly true among emerging nations, which have often pursued mega-events to enhance their reputation or national brand on a global platform:

> Many emerging nations have risked a great deal in betting that hosting of a mega-event can be a fast-track to world recognition and reputation enhancement, and there is considerable evidence that this bet has payoffs in positive impacts on country images and reputations as producers of products and as tourism destinations. (Heslop *et al.*, 2013: 13)

In the case of South Africa, the hosting of sport mega-events has been part of a larger national agenda for nation building as well as showcasing the state as a 'global middle power' (Cornelissen, 2008: 486). The country has staged a series of major and mega-sport events that include the 1995 Rugby World Cup, the 1996 African Cup of Nations (football) and the 2003 ICC Cricket World Cup. However, the 2010 FIFA World Cup will serve as the focus of this case study given its scale and international profile. The CEO of the organising committee clearly stated that the event was: 'about nation-building, it's about infrastructure improvement, it's about country branding, it's about repositioning, it's about improving the image of our country, and it's about tourism promotion' (Allmers & Maennig, 2009: 500). Key development issues associated with South Africa's hosting of the FIFA World Cup are highlighted next based on an empirical assessment of the event (Knott, 2015).

Urban development

The stakeholders' experiences confirmed that this mega-event's perceived success and positive legacy were in large part attributed to the event-linked urban development projects, especially those aimed at improving crucial areas of urban and tourism infrastructure. In addition to the stadiums, private and public entities in the host cities and provinces invested heavily in building facilities and infrastructure as part of associated urban rejuvenation projects. For example, significant airport upgrades were made at the Johannesburg and Cape Town international airports, while a completely new international airport was built for Durban. Other major and minor renovations to domestic terminals were made as part of the event preparations around the country. Additional transportation projects included the 'Gautrain' (a sophisticated new rail service in the Johannesburg and Pretoria urban area), the launch of a new Bus Rapid Transit (BRT) transportation network and the upgrading of the main rail terminal in the city centre of Cape Town.

Authenticity and cultural showcasing

The design, aesthetic and iconographic elements of the new stadiums were consciously developed to reflect local culture and identities. The new stadiums featured eye-catching iconic designs that stakeholders believed enhanced rather than detracted from the cities' skylines. Authentic African themes were used in stadiums such as the Mbombela Stadium in Nelspruit (featuring giraffe-resembling supporting structures and zebra-striped spectator seating) and the Soccer City stadium in Soweto, Johannesburg (modelled on an African *calabash*, a type of traditional wooden bowl). While these designs were used to differentiate the brand of each stadium, as a whole, they distinguished South Africa as a FIFA World Cup host and as an emerging economy with a rich cultural heritage.

Sustaining legacies

Despite lofty objectives and major investments, several of the stadiums built for the 2010 FIFA World Cup have struggled. For example, the Cape Town stadium has been underutilised and is no longer considered a sustainable legacy. Generally, these stadiums have been criticised for not being sustainable post-event despite their innovative designs. Post-event assessment suggests that more effective leveraging of stadiums was required such as being designed for multipurpose use and commercial activities. The stadium in Durban has been successful in this regard with a bungee jump, retail and restaurant attractions and the post-FIFA hosting of a variety of other sport and charity-linked events.

Beyond this, the stadiums should also have been leveraged as key domestic and international tourist attractions in their own right, forming part of the destination branding mix of the host cities. One study respondent emphasised the challenge of leveraging the stadiums, urging South Africans to 'think outside the box' with regard to stadium activities. An important lesson learned in South Africa is that FIFA World Cup legacy projects are not just about impressing a world audience. They need to be planned within a sustainable development

(Continued)

Case Study 4.1: (Continued)

context, aimed at assisting economic development, social cohesion and environmental integrity in the long term. For example, stakeholders cited Barcelona as a city that utilised the 1992 Olympic Games platform for developments that assisted with the brand image transformation of the city and the revitalisation of its tourism appeal. Sustainable development in a country like South Africa requires that the benefits or legacies of such mega-events should reach beyond the immediate event-hosting area or city. Smaller municipal areas in South Africa such as Mbombela would have benefited significantly from infrastructure development projects associated with the new stadiums in terms of road and transport improvements and other tourism services developments.

These projects and activities are examples of the key leveraging efforts that added to the perceived success of the event, enhanced the nation's international reputation and improved tourists' experience, thus positively impacting the nation's brand development. Investments in infrastructure developments at similar events have the potential to provide an enabling environment not only for the delivery of the event itself, but also for the host community, its businesses, tourists and investors, and possibly for the hosting of future large-scale events. If planned, built and managed in a sustainable manner, these initiatives can create positive leveraging opportunities for tourism development. These leveraging activities should be considered within a sustainable development framework that considers the nation's strategic developmental objectives. Although this case study focuses on leveraging opportunities associated with a mega-event, there was support among study respondents for more sustainable smaller, regular, local (home-grown) sporting events. These smaller events were perceived to be less costly and to provide more niche benefits than those normally leveraged from mega-events. Overall, this case study provides evidence in support of the assertions that sport tourism events can be leveraged to build place identity and to position destinations (Higham & Hinch, 2009). While the 2010 FIFA World Cup was deemed a success in terms of positioning South Africa on the world stage during the competition (Roche, 2000), its economic and social legacy has been more problematic at a local and domestic level.

Key Reading

Knott, B. (2015) The strategic contribution of sport mega-events to nation branding: The case of South Africa and the 2010 FIFA World Cup. Unpublished PhD thesis, Bournemouth University.

Additional references cited in this case study are included in the reference list.

One of the dominant interdependencies that characterises globalisation is the relationship between economics and culture. On the one hand, globalisation can be described as a culture of consumption inclusive of sport and tourism. On the other hand, it can be characterised as the consumption of culture (Higham & Hinch, 2009). For example, media conglomerates and major sporting goods manufacturers/retailers like Nike have been identified as powerful forces influencing the globalisation of sport (Harvey *et al.*, 1996; Thibault, 2009). While Maguire (1994) has argued that the influence of developments such as the internet demonstrate that the globalisation process is not necessarily guided or planned, others like Thibault (2009) have demonstrated that major international corporations have been very strategic in their manipulation of global forces. The actions of these players in globalisation are driven by self-interest through the commodification of sport.

Mowforth and Munt (2015) have suggested a similar dynamic in the case of tourism. There is a constant process of commodification driven by the tourism industry's relentless search for 'new destinations'. As these places are increasingly connected to tourism-generating regions, globalisation through tourism is manifest in a very tangible fashion. One example of this is the increasing fragmentation of production through practices such as 'outsourcing' in the production of tourist experiences (Nowak *et al.*, 2009). Other characteristics of globalisation that have been identified in a tourism context include transnational ownership, labour mobility, cross-border marketing and the sale of intellectual know-how (Hjalager, 2007). Even the promotion of destinations in a more globalised world has created challenges through the separation of place image from culture and identity (Morgan, 2014).

Tourism is, in fact, a significant force in the globalisation of sport. Jackson and Andrews (1999) have pointed out the influence of tourism-savvy Disney Corporation on the business model adopted by David Stern during his tenure as the commissioner of the National Basketball League. In her critique of globalisation in sport, Thibault (2009) describes its 'inconvenient truths'. These include the emerging division of labour such as the use of cheap workers from developing countries to manufacture sporting goods, increased mobility of elite athletes from poor to wealthy countries, increased influence by the global media and increased environmental impacts that are in part due to sport-based travel.

One of the most interesting aspects of these debates from a tourism perspective is whether globalisation is leading towards a homogenisation of sport culture or whether local resistance will retain or even foster greater differences between places (Go, 2004; Maguire, 2002). This is especially significant because in a world where there are growing similarities between many places, there really is little need to travel. Homogenisation is seen as a significant concern in tourism (Dwyer, 2014) but has also been raised in the context of sport. For example, Silk and Andrews (2001) suggest that electronic spaces or the 'space of flows' may supersede the 'space of places'. In the context of sport tourism, if sporting culture were to evolve into a homogeneous global culture, much of the existing incentive to travel for sport would be lost. This debate is, therefore, an important one for sport tourism development.

The essence of the homogenisation/heterogenisation debate in sport is nicely summarised by Silk and Jackson (2000):

> [h]omogenisation heralds the advent of an era dominated by creeping global standardization. Heterogenisation, however, rejects the influence of global technologies and products in favour of stressing the inherent uniqueness of localities. The former category suggests that we are becoming more alike and heading towards a uniform global culture. The latter emphasizes cultural differences and the power of the particular. (Silk & Jackson, 2000: 102)

Those who see the forces of globalisation leading to the homogenisation of sport have argued that there is ample evidence to suggest that sport culture is growing more similar throughout the world. For example, the emergence of homogeneous sportscapes as manifest in standardised stadiums and sports fields (Bale, 1993) is consistent with this view. Rowe and Lawrence (1996: 10) also suggest that cultural commentators find 'evidence' of this new phenomenon in international sports media spectacles (such as the Olympic Games and the FIFA World Cup); geographically 'mobile' sports (such as basketball and golf); and US-originated advertising, promotion, marketing and 'packaging' practices (such as celebrity endorsements and the high-pressure sale of sports paraphernalia).

Yet, even commentators who show some level of support for this hypothesis recognise that there are other forces at work that seem to counter the processes of homogenisation.

> Globalization is accordingly best understood as a balance and blend between diminishing contrasts and increasing varieties, a commingling of cultures and attempts by more established groups to control and regulate access to global flows. Global sport development can be understood in the same terms: that is, in the late twentieth century we are witnessing the globalization of sports and the increasing diversification of sport cultures. (Maguire, 1999: 213)

Many sport theorists share the view of co-existent forces for homogeneity and diversification (Denham, 2004; Melnic & Jackson, 2002; Washington & Karen, 2001). At an empirical level, there is also support for the thesis that local resistance ensures that there is a significant degree of difference between local sport cultures. Despite global trends to export US sports like American football, regionally prominent sports like Australian Rules football and rugby league have prevailed in Australia (Rowe *et al.*, 1994). Similarly, it has been argued that New Zealanders have negotiated the introduction of basketball on their own terms (Jackson & Andrews, 1999). Likewise, Bernstein (2000) found that despite the powerful forces of the global media, press coverage of the 1992 Olympics in Barcelona was characterised by local or national perspectives. Nationalistic interpretations of sporting events and performances still dominate media coverage.

Notwithstanding the powerful momentum of globalisation processes, the environment in which they occur is dynamic. This is particularly the case in terms of the political environment, which has seen popular protests against the perception of a neoliberal agenda that favours the concentration of wealth in the hands of fewer and fewer people (Mitlin *et al.*, 2007). More recently, popular movements in Europe and the United States have challenged free trade agreements, and liberal policies on immigration suggest that the processes of globalisation are not inevitable as they recently appeared.

Organisational fragmentation

Globalisation is about increasing interconnections. These interconnections take the form of networks of interacting and interdependent actors. March and Wilkinson (2009: 461) describe them as 'complex and mutable entities that develop and evolve over time in response to environmental and organisational developments and demands'. Pressures to collaborate include global trends such as: (1) the pursuit of public sector effectiveness, (2) public sector budget cuts and restructuring, (3) fragmentation within the sport and tourism sectors, (4) pursuit of best practices and industry trends and (5) institutionalisation within the sport and tourism sectors (Zapata Campos, 2014). The tourism sector is full of variations of these partnerships including tourist boards, industry associations, chambers of commerce, convention bureaux and strategic alliances, while sport features a more hierarchical public sector framework. Sport tourism alliances such as those found in Canada and in Japan (Focus Point 4.2) are increasingly being introduced as a way to capture the synergies related to sport events.

Focus Point 4.2: Sport Tourism Alliances in Canada and Japan

Initiated in 2000, the Canadian Sport Tourism Alliance (CSTA, 2000) has grown from 18 founding members to over 500 members including 200+ municipalities, 250+ national and provincial sport organisations, 20+ educational institutions and a variety of product and service

(Continued)

Focus Point 4.2: (Continued)

suppliers to the industry. It has worked in close cooperation with Destination Canada to support the development of event-based sport tourism in Canada. The key to the CSTA's success has been its ability to connect the various stakeholders in this realm and to provide them with information and resources. Sport organisations are effectively partnered with host destinations to bid on and deliver a wide range of sporting events. The CSTA bases its activities on a solid planning framework that is guided by its mission to increase Canadian capacity and competitiveness to attract and host sport events. Specific objectives include: (1) to establish Canada as a preferred sport tourism destination; (2) to enhance the image and profile of the sport tourism industry; (3) to facilitate networking, educational and communications opportunities; (4) to develop and facilitate access to industry tools; (5) to build investment in sport tourism from the public and private sectors; and (6) to coordinate research, data collection, monitoring and reporting of activity within the sport tourism industry. The success of the CSTA is one of the reasons that the Japan Sport Tourism Alliance (JSTA, 2011) used CSTA as a model for its own development in 2011. However, the JSTA has made some substantial additions to the CSTA model. It has increased the scope of its mandate to include participant events and active sport. This expansion of its mandate is more in line with the full range of sport tourism attractions (see Chapter 2) and as such, it increases the JSTA's ability to embrace a broad range of sport tourism strategies in support of desired outcomes like regional development.

Key Reading

Canadian Sport Tourism Alliance (2002) CSTA official website. See http://www.canadiansporttourism.com/portal_e.aspx (accessed on 10 August 2017).

Japan Sport Tourism Alliance (2011) JSTA official website. See http:// sporttourism.or.jp/ (accessed on 10 August 2017).

It is challenging to articulate goals and implement successful strategies in such a complex environment. Organisational fragmentation therefore remains an issue. The broad range of interests found within many of these partnerships presents substantial challenges despite the common ground that may exist (Zapata Campos, 2014). A prerequisite to addressing these issues is not only recognition of existing and potential networks, but a genuine willingness and ability for these players to work together in pursuit of their common interests related to sustainable sport tourism development. This outcome is most likely to occur when each partner has a clear understanding of the synergy that can be captured by the partnership and how this synergy makes a contribution to each partners' core mandate.

Until recently, however, partnerships in sport tourism have been largely ineffective. For example, a study of six European states in the early 1980s identified a perceived linkage between sport and tourism in the minds of participants, commercial providers and local authorities (Glyptis, 1991). However, despite this perception, there was a lack of conscious integration by policymakers, planners and public providers at the national level. Weed and Bull (1997a) noted that by the late 1990s there were still very few joint sport tourism initiatives among the regional agencies responsible for sport (Sports Council regional offices) and tourism (regional tourist boards) in England. Even in Australia, which is a leader in sport tourism partnerships, the primary rationale for the development of its policy on sport tourism was the lack of an identity and cohesiveness perceived in this area (Commonwealth Department of Industry, Science and Resources, 2000).

The need for partnerships and strategic alliances in tourism and in sport has grown over the past 25 years, especially as government resources that were traditionally directed towards these areas have been reduced (Zapata Campos, 2014). While a broad range of models for partnership have emerged, they can generally be described as '...a voluntary pooling of resources (labour, money, information, etc.) between two or more parties to accomplish collaborative goals' (Selin & Chavez, 1995: 845). Partnership is particularly important in a sport tourism context given the many stakeholders involved. In his study of the restructuring of winter resorts in the French Alps, Tuppen (2000: 337) noted that development '...results from the actions of different organisations and interest groups in both the public and private sectors, often rendering management a complex task'. In addition to the sheer number and diversity of stakeholders who may be involved, the dynamic of power shifts among these stakeholders throughout the temporal course of development further complicates these partnerships.

There are various additional constraints to sport tourism partnerships. Competition between stakeholders, bureaucratic inertia and geographic as well as organisational fragmentation are typical (Selin & Chavez, 1995). In the specific context of sport tourism, the Commonwealth Department of Industry, Science and Resources (2000) highlighted the lack of awareness of the mutual benefits of establishing alliances and difficulties in coordinating resources and information. Based on a study of sport and tourism policy in the UK, Weed (2003) added that the main constraints to successful sport tourism partnerships were ideological differences, inconsistent understanding of key concepts and definitions, differing regional contexts, unsupportive government policy, intransigent organisational cultures and structures, and individuals with contradictory styles and interests.

The bottom line is that successful partnerships need partners who recognise their mutual interest. Participating organisations must also be characterised by a domain focus that is goal oriented rather than organisation oriented. Kennelly and Toohey's (2014) case study of a strategic alliance between (1) an Australian national sport organisation, (2) Australian rugby union and (3) sport tour operators suggested that these sport tourism alliances have potential. They found that alliance partners accrued a range of tangible (including financial) and intangible benefits that encouraged continued support for the partnership. Moreover, the partnership also benefited rugby fans by providing a better rugby experience and served the public interest more broadly by making a positive contribution to the national economy.

Conclusion

The underlying premise of this book is that the sustainability of sport tourism can be facilitated by active intervention in the processes operating in the realms of sport and tourism. Such interventions should target the spectator event, participant event, active and heritage dimensions of sport tourism. Planning will help to optimise the design (or redesign) and development of sport tourism products and services and by so doing, will impact the environment in which sport tourism functions.

Sport tourism developers need to be conscious of the challenges that they face in terms of commodification/authenticity, globalisation and organisational fragmentation. They should, however, also be aware of the opportunities that accompany these issues. In terms of commodification and authenticity, the protection of the sport attraction should be a fundamental objective. Care must be taken to keep the spirit of sport competition and the entertainment spectacle in appropriate balance. Media representations need to be rooted in place (Chapter 6). The authenticity of the sport attraction should be retained without suppressing the dynamic evolution of a sport (Chapter 10). Relative to many other types of tourist attractions, sport has a major advantage in terms of the joys of performance, its unpredictable drama, its embodied nature and the communities that are associated with it. By maintaining the integrity of sports, spectators and active sport tourists will have access to the 'backstage' of sporting destinations where they are more likely to enjoy authentic experiences of place.

Globalisation issues play out around the trade-off between global and local interests. In many cases, the motivation for hosting major sporting events is to establish the host city as a significant player in a global context.

For this to happen, the local context of the event needs to be considered and negotiated. Active and heritage-based sport tourism attractions can be used to mediate global demands. Sport tourism strategies to foster a positive local identity and destination image should be driven by significant local input.

Sport tourism partnerships should be established and operationalised in a way that is mutually beneficial to partner members and the wider society (from the local to the global scale). One of the first steps in such an exercise is to articulate the advantages and goals of cooperative partnerships which usually require careful negotiation. It is particularly important for the tourism industry to demonstrate the benefits of involvement for sport groups. Sporting interests must be convinced that their cooperation will result in increased gate receipts, facility development, new participants for their sport and similar types of benefits. It is not sufficient for the tourism industry to be a silent partner, and significant beneficiary, of sport tourism. Beyond the recognition of these benefits, specific strategies need to be developed to address the constraints on alliances and partnerships that are discussed in this chapter. Unproductive competition between stakeholders is a significant barrier to development. Communication linkages that reduce fragmentation between and within stakeholder groups are one way of addressing these barriers.

Part 3
Sport Tourism Development and Space

5 Space: Location and Travel Flows

Although sport tourism can boost export spending in a defined region,
not all communities have an equal likelihood of successfully
hosting an event, tournament, or team.
Daniels, 2007: 333

Introduction

Sport tourism development takes place within a complexity of spatial parameters. Different sports are reliant to differing degrees on natural or built resources or a combination of natural resources and built facilities. Some sports are rigidly anchored to specific and non-transportable natural resources. Others are relatively free of resource constraints and may be located where proximity to concentrations of population and/or a tourism economy offer the greatest competitive advantage (Bale, 1989; Mason & Duquette, 2008). Distance–time–cost thresholds also shape the spatial travel patterns of sport tourists (Bale, 1989). However, sport tourism market range and travel flows can be moderated and shaped by strategic planning actions and partnerships at a range of spatial levels (Higham, 2005). Successful strategies require that consideration be given to the relationships that exist between sport, tourism and space. This chapter discusses the locations where different forms of sport tourism take place, and the 'movement of tourists from originating markets to leisure destinations of their choice' (Mitchell & Murphy, 1991: 57). It examines sport tourism resource requirements, destination hierarchies and travel flows. The locational requirements and travel flows associated with sports that take place in central and peripheral areas are then addressed, followed by consideration of the spatial travel patterns associated with active sport tourism. The concept of scale, so central to geographical theory (Higham & Hinch, 2006), and how it is being reshaped by global political and environmental change, is a recurring element of these discussions.

Sport, Tourism and Space

Space and place are concepts that are central to the geography of sport (Bale, 1989) and the geography of tourism (Hall & Page, 2014; Lew, 2001; Pearce, 1987). Unlike recreation and free play, many sports require defined spatial delineations, such as the length of a marathon (26 miles, 385 yards) or the physical parameters of a basketball court (29 × 15 m). The spatial boundaries that are applied to sports are written into rules and codes of regulation. These rules may be explicit in terms of player movement, as in the case of netball (where, for example, defenders must remain in the defensive half of the court), or implicit, where a defensive formation will be weakened or broken if a player moves out of position. 'In many cases sport involves the dominance of territory or the mastery of distance; spatial infractions are punished and spatial progress is often a major objective' (Bale, 1989: 12).

Tourism is also characterised by a spatial component. To be considered a tourist, individuals must leave and then eventually return to their home (Chapter 2). Travel is one of the necessary conditions of tourism, and it is for this reason that the spatial implications of tourism are important. A variety of qualifiers have been placed on this dimension including a range of minimum travel distances, but the fundamental concept of travel is universal (Hall, 2004). The spatial dimensions of sport and tourism are critically important to the respective fields.

The spatial element of sport tourism, as it is addressed in this chapter, centres on the locations and regions in which specific sports take place, travel flows that are directly or indirectly associated with those sports and the ways that these flows may be moderated and facilitated. A variety of questions emerge from this discussion associated with the resource base, location and management of sport attractions. For example, to what extent can sports resources be reproduced and transported? Similarly, what are the implications of changes within a sport in terms of the propensity of spectators to travel to attend a sporting event? And at what scales of analysis, ranging from local/regional to global, do particular sport and tourist flows interplay?

Spatial analysis of sport tourism

The spatial analysis of sport tourism involves the study of the locations in which sports occur, and the movement of tourists into and at these locations. Such an analysis finds its theoretical foundation in the geography of sport (Bale, 1989, 1993; Rooney, 1988), which introduces concepts such as central place theory, distance decay and location hierarchy for consideration in the study of sport tourism. This analysis also draws on the geography of tourism, which considers the 'spatial expression of tourism as a physical activity, focusing on both tourist-generating and tourist-receiving areas as well as the links between' (Boniface & Cooper, 1994). The concept of scale, from city and state/province to national and global scales of analysis, is critical to these discussions. So too, interestingly, is the growth of e-sports and virtual reality, which opens the way to experiences of sport that are largely or entirely free of (physical) spatial parameters.

The (physical) spatial elements of sport tourism vary dramatically between sports that tend to be centrally located (urban) and those that take place in peripheral regions. Once again, definitions of central (core) and peripheral (periphery) change with scale or resolution of analysis, from the globally peripheral to the national and regional peripheries. The spatial elements of sport tourism in central and peripheral locations justify separate analysis. Sport tourism in urban locations is typically based on constructed sports facilities, whereas sport tourism in peripheral areas is generally based on the presence of nature-based resources, which may be modified or complemented by built facilities. This distinction reflects Boniface and Cooper's (1994) tripartite classification of tourism resources as:

(1) User orientated: Centrally located, intensive developments providing proximity to markets and tourism infrastructure, and based on built or artificial facilities and attractions.
(2) Intermediate: Located with a view to accessibility, based on resources that are built and/or natural.
(3) Resource based: High quality natural resources that tend to be spatially removed from centres of population and located on the basis of remote and limited resource availability.

The spatial concept of distance decay applies to both sport and tourism. For example, in the case of elite sport, a discernible pattern exists in terms of the home or away status of a sports contest, and the probability of winning. Not only is winning away less probable than at home, but 'the probability of winning forms a clear gradient according to distance from home' (Bale, 1989: 31). The further a team travels from its home venue, the less likely it is to win.

In the context of sport tourism, sports that take place in central locations are generally advantaged by proximity to markets. Residents of adjacent regions or from peripheral areas are less likely to travel to a sporting event or activity than those located nearby (Daniels, 2007; Pearce, 1989). The gravity model of distance decay suggests that tourist flows decrease with distance from the origin (Boniface & Cooper, 1994). In theory, therefore, the power of attraction that a sport may exert upon the travel decision process diminishes as the distance between the origin and sport site or venue increases. The distance decay function that underlies the gravity model has been understood to be influenced by increasing travel costs and declining knowledge of distant locations (Mitchell & Murphy, 1991). Therefore, in the case of event sport tourism, all other things being equal, the further a team travels to compete the less likely that their home-based supporters will accompany them.

In reality, a linear distance decay function is moderated by a range of factors (Miossec, 1977), such as cultural and climatic characteristics, which may act as barriers or facilitators to travel (Cooper *et al.*, 1993; Mitchell & Murphy, 1991). Travel flows may be mediated by a number of interrelated variables (Boniface & Cooper, 1994). Zonal travel patterns can be 'modified by the hierarchy of resort destinations, the spatial advantages offered by major transport routes, and locations with outstanding (unique) reputations' (Mitchell & Murphy, 1991: 63). The provision of low-cost (budget) air services over the last decade is one obvious example of infrastructure and service development that may mediate physical distance and redefine space (Casey, 2010). The distance decay function of sport tourism may also be mediated by such things as the quality of the opposition, importance of the competition, duration of the event, number of matches that will be played by a team or athlete and the travel distances between competition venues. The attractiveness of the sport and the location (destination) where the sport takes place may in combination create a powerful tourism proposition (Higham, 2005), which can be further embellished through strategic branding, packaging, leveraging and bundling initiatives (Chalip, 2006; Chalip & Costa, 2005; Chalip & McGuirty, 2004; O'Brien, 2007). These factors, in combination, can significantly enhance the spatial range and travel flows associated with sport phenomena.

Sport locations, location hierarchies and tourism

Modern sports exist in a continual state of change. The dynamics of change are often driven by economic processes that bear upon the structure of competitive sports (e.g. the development of new league competitions), the location of sport facilities, the changing distribution of nature-based sports resources and the rise and fall of sport attractions. Bale (1989: 77) refers to 'the growth and decline in importance of different sport locations', which parallels Butler's (1980, 2006) tourist area life cycle theory (Chapter 10). These dynamics have implications for the scale of the player and spectator catchments. Within the ranks of professional sports, the limitations associated with only drawing players from areas nearby the home team site are alleviated through external recruitment (regional, national or global), player transfers and draft schemes. The spectator catchment, and the propensity for residents and non-residents in different regions to attend live sport, is a separate but similar issue that is of particular relevance to sports marketing managers. Professional sports franchises seek to attract the best athletes to their players' roster and, simultaneously, build and retain the widest possible catchment of fans. Interests in building a team profile and supporter base beyond the geographical area that the team actually represents may be significantly advanced in collaboration with tourism destination managers at sport locations.

Central place theory lends itself very conveniently to the study of sport and tourism (Daniels, 2007). Sports attractions exist within a hierarchical organisational structure (Table 5.1) in a similar fashion to other tourist attractions (Leiper, 1990). This hierarchy reflects the fact that some sports centres primarily draw upon a local catchment, while others situated higher in the sports hierarchy draw upon regional, national and international catchments. Bale (1989: 79) explains that sports facilities situated in central locations are located 'as close to potential users as possible in order to maximise pleasure from the sport experience and to minimise travel, and hence cost'. This characteristic has been complicated in recent years as new factors have emerged that influence the status of sports locations. These factors include facility sharing, changing access to infrastructure and travel nodes, proximity to tourism and service developments, and prominence in media markets (Stevens, 2001).

The notion that the demand for sports decreases with distance from the location at which the sport is consumed applies to sport spectatorship as well as participation. Bale (1989) introduces the term 'spheres of influence', which describes the power of attraction that sports teams exert upon spectators. As noted in references to the model of distance decay, the slope and range of the spatial demand curve for sport spectatorship is elastic. It may be influenced by factors that are, to some degree, beyond the immediate control of sports managers, such as the fortunes of the team (win/loss record), circumstances of the competition, league position and weather. A distinct range of factors influence the demand curve relating to sport participation. These factors include costs of access (White & Wilson, 1999), the standard of the sport resource and the uniqueness of the sport tourism experience. Additional touristic opportunities, such as visiting friends and relatives, or achieving other desired tourist experiences at a destination may also moderate the demand curve.

The concept of the sport location hierarchy is central to these discussions. Not only do sports resources exist in a location hierarchy (Bale, 1989), but

Table 5.1 Theory of sports locations

(1) The main function of sports locations is to provide sports outlets for a surrounding hinterland. Sports facilities are therefore centrally located within their market areas.
(2) The greater the number of sports provided, the higher the order of the sports location.
(3) Low-order sports locations provide sporting facilities that are used by smaller catchment areas; the threshold population needed for the viability of a lower-order place is smaller.
(4) Higher-order locations are fewer in number and are more widely spaced. They have large population thresholds.
(5) A hierarchy of sports locations exists in order to make as efficient as possible the arrangement of sports opportunities for (a) consumers who wish to minimise their travel to obtain the sport they want and (b) producers of sports who must maintain a minimum threshold of customers to survive.

Source: After Bale (1989).

so too do tourism destinations (Higham & Hinch, 2006). These hierarchies are competitive and dynamic such that while sports may evolve in terms of participation and interest (Coakley, 2017), tourism destinations may rise and fall in prominence (Butler, 1980) (see also Chapter 10). As such, sport and tourism managers may collaborate to bring together powerful sports experiences, historically significant moments and unique destinations in a way that encourages travel decision-making. The key, according to Harrison-Hill and Chalip (2005), is to create a powerful synergy between the sport and the destination. Destinations that have succeeded in creating this synergy are likely to be more prominent than competing destinations, and therefore ascend the sport tourism location hierarchy. The stadium location hierarchy is a particularly competitive aspect of sport tourism (see Focus Point 5.1).

Focus Point 5.1: The Stadium Location Hierarchy

Sports stadiums exist within a location hierarchy that is dynamic and competitive (Bale, 1989). The status of a stadium within the location hierarchy has important implications for city economic development, tourism and service sector interests. Millennium Stadium (Cardiff), Rogers Centre (Toronto), Etihad Stadium (Melbourne) and Westpac Stadium (Wellington) are examples of stadiums that have been developed in inner-city sites, with immediate proximity to central transport nodes and the urban services sector. The concept of the location hierarchy confirms that beyond the importance of urban stadium facilities, it is the relative position of a stadium within a spatio-political region that is of considerable importance. After 50 years of developmental status quo, a period of intense stadium (re)development has occurred in New Zealand in recent years, with implications for hosting tier one sports events, the location of professional sports franchises, attracting concerts and other cultural events, attention in media markets and sponsorship. Triggered initially by the professionalisation of rugby union in 1996 and the subsequent hosting of the Rugby World Cup (RWC) in 2011, stadiums throughout New Zealand have been subject to a period of intense redevelopment. Most notably, Wellington's historic Athletic Park was retired in the late 1990s, and replaced by the new Westpac Trust Stadium in downtown Wellington in 1999. This anchor project (among others) repositioned Wellington as a tourism destination and, with the successful bidding to host events such as the IRB Wellington Sevens series (1999–2017), propelled Wellington to the top of the stadium location hierarchy in New Zealand. Subsequent stadium development in the southern New Zealand city of Dunedin offers further insights into the stadium location hierarchy. The Forsyth

Barr Stadium (Dunedin, New Zealand) was a greenfields project that resulted in the retirement of the historic Carisbrook Stadium, and the development of a new stadium in an inner-city location adjacent to the Dunedin railway station and the University of Otago. Completed in 2011 with a fixed roof, it features a number of revolutionary design features that underpin its claim to be the world's only fixed roof (non-retractable) stadium with a permanent natural turf playing surface. The roof is made of ethylene tetrafluoroethylene (ETFE), a fluorine-based plastic that was originally developed for the space industry and subsequently used in the construction of the Eden Project (UK) and Beijing's 2008 Olympic National Aquatics Centre (the Water Cube). This technology affords protection from the elements but allows ultraviolet light to support natural grass growth irrigated by water that is harvested from the stadium roof. The fixed roof is considered a strategic advantage of great importance in attracting sport and other events, and boosting advanced ticket sales because the vagaries of the southern New Zealand climate are neutralised and eliminated from the purchase decision. It affords a spectator experience that is superior to any other stadium in New Zealand. At the same time, the question arises as to whether the production of sports in standardised facilities that represent a complete removal from the unique elements of place actually represents a threat to tourism. The stadium roof represents a loss of authenticity for some because the weather was once an important aspect of playing style that offered a unique home ground advantage. Will there be a backlash? Does standardisation ultimately result in the loss of local uniqueness, or does it actually enhance place differentiation, uniqueness and recognisability? Currently, the fixed stadium roof is unique, but it is likely to be replicated perhaps most immediately in the neighbouring city of Christchurch, which is being rebuilt after the 2010–2011 earthquake sequence that destroyed Christchurch's Jade Stadium (formerly Lancaster Park), which had been extensively rebuilt immediately prior to the Christchurch earthquakes. If so, the status of unique destinations that seek to host sports events may be compromised as the venues for sport become more standardised and less distinctive and recognisable, but the dynamics of the stadium location hierarchy will continue.

Key Reading

Bale, J. (1989) *Sports Geography*. London: E&FN Spon.

Forsyth Barr Stadium. Construction facts and figures. See https://forsythbarrstadium.co.nz/news/article/716 (accessed 8 September 2017).

The Spatial Analysis of Sport Tourism in Central Locations

In practice 'a vast number of physical, economic and social barriers will contribute to a distortion of the central place model' (Bale, 1989: 81; see also Daniels, 2007; Mason *et al.*, 2008). For example, the catchment population required to support a professional sports franchise will vary as determined by the propensity of residents within the catchment area to support the team. Small city teams, such as the Saskatchewan Roughriders (Canadian Football League) and the Green Bay Packers (National Football League, USA), serve to illustrate that the level of support a team receives at its stadium may bear little resemblance to the population of the host city. The Roughriders, for instance, survive due to a strong team following from across the province of Saskatchewan. This example confirms that 'human and cultural factors can upset the rationally economic world predicted by central place models' (Bale, 1989: 82). These factors help to explain the higher than expected loyalty from within and beyond the local spectator catchment that some teams are able to generate.

Inter-urban travel is an increasingly common by-product of sport. Bale (1989: 112) notes that 'in an age of relatively easy inter-regional and international travel, sports events are able to generate substantial recurrent gatherings of peoples ... and hence ... [they contribute] ... to the wealth and economic dominance of the big city'. Increasingly, sport events and facilities are being used as economic anchors in the entertainment districts of higher-order urban centres (Mason *et al.*, 2008). The sports location hierarchy and the spatial demand curve, both of which are subject to change over time, influence the status of a sport tourism destination. In most cases, sports teams compete at a home venue to which spectators will travel from within the host region and, possibly, further afield (Gibson, 2002; Higham & Hinch, 2000).

Traditionally competition leagues are characterised by a series of home and away games, which are attended by supporters of both teams in varying proportions. Home games tend to be dominated by hometown supporters, although sport and tourism managers may devise strategies to encourage visiting fans, casual spectators and tourists at a destination to attend a sports contest. It is also noteworthy that home team supporters are not necessarily hometown residents. As an example, since the return of Newcastle United football club to the English Premier League, an estimated 170,000 long-weekend trips to Newcastle per year have been undertaken by Norwegians who support this football club (Law, 2002). Strong and enduring support for Newcastle and Liverpool football clubs in Norway arises from the initiation of regular television coverage of English football in Norway in the 1970s, when these clubs were forces of the English first division (now Premier League) (see also Case Study 6.1).

Other models exist in the spatial organisation of sport competitions. 'An alternative form of spatial organisation is for the sport to travel to the people in order to attract sufficient business to meet its threshold population' (Bale, 1989: 85). While some sports are rigidly anchored to specific and non-transportable natural resources, others are relatively free of resource constraints and may be transported (Chapter 7). A marathon course, for example, can be located to take advantage of concentrations of population, distinctive urban landmarks or unique scenic settings.

Professional sports tours involve a tour circuit incorporating a recurring seasonal calendar of events in different venues where competition takes place. These tours are designed to improve spectator access to sports and therefore revenues, and are scheduled in cases such as golf (e.g. Professional Golf Association Tour) and tennis (e.g. Association of Tennis Professionals Tour) to take advantage of seasonal conditions at the destination (see also Chapter 9). A professional tour has two noteworthy implications for sport tourism. First, it transforms the athlete or contestant into an elite sport tourist as the tour circuit moves from one venue, city, country or continent to the next. Secondly, it creates the opportunity for sport tourism development and periodic marketing associated with the regularly recurring visit of the professional sports tour. World Rugby's (formerly the International Rugby Board) HSBC World Seven Series, most notably the annual tournament that is hosted in Hong Kong, has been developed and promoted as a sports festival, often in association with other urban tourist activities. While the Hong Kong Sevens has achieved sustained success for several decades, the Wellington (New Zealand) Sevens tournament, which in the 1990s developed into a colourful festival and platform of contemporary culture, has fallen in prominence in recent years. This has been attributed to the emergence of the rival Auckland Nines (Rugby League) and the association of the Wellington Sevens with excessive alcohol consumption (Gee, 2014).

Formula One, which currently involves 17 races in the annual Grand Prix circuit, is another example of an annual professional sport competition that has been developed on the principle of periodic marketing, although this sport is characterised by intense competition between cities to host F1 Grand Prix events (Henderson et al., 2010). Recurring sports events, despite the threat of locational change, stand in contrast to biennial and quadrennial sports events such as the Federation Internationale de Football Association (FIFA) Football World Cup and the International Cricket Council (ICC) Cricket World Cup, the International Amateur Athletics Federation (IAAF) World Championships and the Olympic Games. The cities that host these events change, as determined by bidding processes, which presents different challenges and opportunities to sport tourism destination managers. In these cases, it is particularly important that sport and tourism managers collaborate to take full advantage of the opportunities that these events present; be they in the form of regeneration, business development, destination marketing, reimaging, tourism or event legacy (Weed, 2007). One such strategy has been to build the imagery associated with events by, among other things, hosting specific sports in association with prominent and iconic tourism attractions at the host destination. Images of Horse Guards Parade and Buckingham Palace (London 2012 Olympic Games), Christ the Redeemer (Rio de Janeiro 2016 Olympics Games) and the Champs-Élysées and Arc de Triomphe (Paris, Tour de France), are classic examples of the use of iconic urban monuments in the projection of sports event imagery to a global audience. The use of landscape vistas and recognisable heritage sites may also be considered to effectively project imagery to external audiences in association with sports events.

Sport tourism market range

The market range of a sports team varies according to a wide array of factors. These include historic record, style of play, team image, public promotion and the success of the team, which influence the status of a team as a tourist attraction (Hinch & Higham, 2001). Similar factors apply to the visiting team as well as the home team. A recent trend in North American professional sport leagues is to charge spectators a premium to watch the top-ranked visiting teams. For most sports clubs, the spectator catchment is local/regional in scale, although some successful clubs have managed to extend the range of their spectator markets, fan base and media reach.

'Hallmark teams' are those that 'regularly attract large spectator crowds (and) have now become synonymous with tourism place promotion as well as short break leisure tourism packages' (Stevens, 2001: 61). In such cases, the scale of analysis that applies to spectator flows may be international or global. Bale (1993) notes that football clubs such as FC Barcelona and Manchester United receive high levels of media attention. This has helped to build a support base that extends far beyond the location that teams actually represent. While Manchester United football club divides the city of Manchester with its local club rival Manchester City, the club is an international sports brand with a global supporter base (Hill & Vincent, 2006), which built in the 1990s/2000s on the multiple brand personalities of David Beckham and particularly his prominence in the Asian markets (Vincent et al., 2009). FC Barcelona now has more official fan club members in Indonesia than it does in Spain. The implications for tourist market range are significant. Manchester United Premier League games played at Old Trafford regularly attract between 4000 and 6000 international tourists to the Greater Manchester area (Stevens, 2001). Similarly, 46% of all spectators that attend Baltimore Orioles (USA) baseball matches at the Camden Yards Stadium are sport excursionists and sport tourists, approximately 11,000 of whom remain in Baltimore for at least one night (Stevens, 2001).

The spatial travel patterns associated with a sports team may be mapped using readily available secondary data such as the places of residence of season ticket holders, or fan club members. Extending the study of sport spectator travel flows to include 'casual' spectator markets is now much more straightforward than in the past due to the development of electronic ticketing and data systems. These big data sources allow insights into attendance over time which may illuminate key factors influencing spatial travel flows and attendance. These analyses may afford some idea of the range and specific regions from which sport tourism spectators originate. Extending market range beyond the local geographical boundaries of a designated team franchise area may be achieved nationally or internationally through match attendance, as well as merchandise sales or supporters club memberships. The continued success of a team influences its market range, but enduring success is very rare. This factor alone cannot explain the sustained and extended fan bases that some teams enjoy. Individual star players and the aura, glamour and heritage associated with teams and the venues at which they compete contribute to the enduring allure of some sports teams. The same factors may influence the propensity of visitors to engage in nostalgia sport tourism (Ramshaw & Gammon, 2016). The atmosphere of the home stadium, the colour and parochialism of the home fans and the public presentation of prominent team players may also bear upon the supporter catchments that are generated by sports teams. These factors apply to sports teams and professional franchises at various levels of competition and scales of analysis (Mason & Duquette, 2008).

It is evident that the spatial range of sport tourism phenomena and related mobilities has expanded considerably (Higham & Hinch, 2009). The driving forces and consequences of globalisation (see Chapter 4) have been the focus of much academic attention in the fields of sport and tourism (e.g. Maguire, 1994, 1999; Mowforth & Munt, 2015; Silk & Andrews, 2001). Higham and Hinch (2009: 10) note that 'sport and tourism have figured prominently in the development of new relationships between cities, regions and states in terms of international trade, business development, capital investment and job growth'. Important debates centre on who wins and who loses under neoliberalism and globalisation (Harvey, 2007), and whether globalisation leads to increasing standardisation or whether local resistance and negotiation will retain or even foster greater differences between places (Bale, 2002; Hall, 1998; Page & Hall, 2003; Silk & Andrews, 2001). While it is generally agreed that forces of and resistance to homogeneity coexist (Harvey & Houle, 1994; Washington & Karen, 2001), the EU Brexit referendum (23 June 2016), Theresa May's triggering of Article 50 (29 March 2017) and the US presidential election (8 November 2016) were historic moments in the context of globalisation. These seismic shifts in global politics will reshape the spatial patterns of sports leagues, travel flows, sport labour markets, sport labour migration and the wider mobilities of sport-related tourism.

Sport, space and the visitor experience

The distances that sport tourists travel usually influence the experiences that are pursued at the destination. The time–distance–cost thresholds of tourism are such that the increasing investment of discretionary time and income on travel will bear upon most aspects of the visitor experience (Chapter 8). For instance, the further sport tourists travel, the more likely it is that they will spend some time at the destination engaging in other types of touristic activities (Nogawa *et al.*, 1996). It is also noteworthy that the area that a sports team represents may in fact require 'home' supporters to travel considerable distances to support their team. A national team performing in international competition at home may attract domestic supporters from throughout the country that it represents, who travel long distances to attend as domestic tourists. Indeed, expatriates may also return to their country of origin to support or compete on sports teams (Higham & Hinch, 2006). Thus, the spatial area that a sports team or club represents may vary considerably from the spatial extent of the team's support catchment. So, for example, Manchester United FC represents one part of the city of Manchester, quite separate from local rivals Manchester City FC; both are highly successful Premier League football clubs with loyal and global supporter bases. This raises the prospect of spectators travelling as domestic or international tourists without feeling that they are leaving 'home', or indeed feeling that they are going 'home', which may have interesting implications for the visitor experience.

Visitor expenditure patterns associated with the sport tourist experience are of particular interest to sport, tourism and service industries. Insights exist into 'both the costs and benefits to a community of attracting a professional sports outfit and the economic impact of an existing sports franchise on the city in which it is located' (Bale, 1993: 77). The expenditures that may be associated with the location of a sports club or franchise in an urban area may include club expenditures, or those associated with the production of the sport, and expenditures generated by local and non-local spectators. Such expenditures will vary with the size of the urban area and the definition of its geographical parameters. Bale (1993: 81) points out, in direct reference to English football, that 'as distance from the football club increases, the positive spill over effect on retailers is likely to decline until a particular point is reached at which the club has no direct economic impact at all'. It is noteworthy that the spending patterns of different sport spectator catchments may be quite unique, with variation between local and non-local visitor expenditure patterns particularly evident (Gibson et al., 2002). However, relatively little research has been committed to this aspect of sport tourism (Chapter 8).

Sport tourism and the status of sport centres

At the local scale, urban centres have been at the forefront of a new phase of entertainment consumption, which Belanger (2000) describes as the spectacularisation of space. This process is 'creating a new urban landscape filled with casinos, megaplex cinemas, themed restaurants, simulation theatres, stadia and sports complexes' (Belanger, 2000: 378). In many cases, this new urban landscape exists in nodal entertainment enclaves that may function as 'sport precincts' and 'tourist precincts' (Leiper, 1990; Mason et al., 2008). 'These group-specific combinations of spatially related attractions and facilities are called "complexes"' (Dietvorst, 1995: 165). The status of sports centres is enhanced when facility developments are planned in coordination with entertainment, tourism and service sector interests. Central sports locations that are situated adjacent to city service and entertainment areas have become an important aspect of the planning for sports centres (Chapin, 2004; Gratton et al., 2005; Thornley, 2002).

The development of the modern stadium features prominently in advancing the status of sport centres that function as tourist destinations. Stadiums have been developed alongside hotel and convention centre complexes as part of urban regeneration and inner-city tourist-based development programmes, and some North American cities have collaborated with sports franchises to maximise the heritage value of tourist experiences by embracing the sporting past in urban sports facilities (Mason et al., 2005). These projects have stimulated the development of the service industry, including travel agents specialising in sport tourism, to accommodate the needs of tourists. These developments, in combination with ancillary tourism services such as accommodation, transport, dining and entertainment, enhance the status of sports centres. In Australia, the Melbourne and Olympic Parks (MOP) precinct is a case example of the development of concentrated and world-class urban sports resources in a manner that is closely tied to tourism and service sector development (Hede & Kellet, 2010; Smith, 2010). Beyond immediate sports facilities, the hosting of sport mega-events can bring significant change to destination image (Kaplanidou & Vogt, 2007; Smith, 2005) and urban form. It is critically important that sport and destination managers, as well as civic leaders and urban planners, are conscious of the potential for sport and event development to drive exclusionary practices in host cities (Case Study 5.1).

Case Study 5.1: Mega Sports Events and the 'Geographies of Exclusion'

Arianne C. Reis, School of Science and Health, Western Sydney University, Australia

Mega sport events such as the Olympic Games have the potential to transform cities. This transformation, however, is not always a positive one, nor is it equally distributed across the host city and all of its residents. Indeed, recent and not so recent experiences have called into question the benefits of event-induced urban transformations. Gustafson (2013: 199), for instance, reminds us of the dramatic infrastructural overhaul carried out in Atlanta in preparation for the 1996 Olympic Games, with the 'destruction of ... large public housing projects to make room for Olympic venues, to *clean up* the city's neighbourhoods around venue sites and to remove the city's homeless population from the emerging Olympic landscape'. Gustafson (2013) estimates 30,000 evictions and displacements between 1990 when the Atlanta Olympic bid was successful and the hosting of the event in 1996. In London, young residents of the Stratford area of Newham, in the East London region where most of the Olympic venues were located, reported the gentrification of their neighbourhood and concern about their ability to remain in the area after the Olympics Games (Kennelly & Watt, 2012). Also in 'Rio 2016', Silvestre and Oliveira (2012) reported on the changing geographies of the west zone of the city and negative effects on the several low socio-economic status (SES) communities that had, at least four years before the event, already been displaced to make way for new highways, metro lines and stadiums. Clearly, these urban transformations have important impacts on the lives of individuals and communities, frequently reinforcing the exclusion of those who are already living on the margins of society (Broudehoux, 2016).

These exclusionary practices have considerable ramifications for different aspects of living in these host cities. One of these aspects is leisure practice, including participation in sport, the very *raison d'être* of these events. It is widely accepted in the academic literature that the physical environment in which one lives influences leisure and physical activity habits (Sallis *et al.*, 2006). Not only can the availability of leisure spaces shape leisure participation but also our relationships with these spaces can facilitate and influence the way we move and use our bodies, and help develop a 'sense of place', fundamental to the construction of individual and community identity (Sampson & Goodrich, 2009). Fragmented, unfamiliar, unsafe, dirty places may have a significant impact not only on how a person feels about where she or he lives but also how she or he feels about herself or himself. Public leisure spaces within communities can thus suggest a lot about the identity of that community, an identity that can certainly be imposed through external forces but that, nonetheless, significantly impacts on how those living within that community see themselves. Previous research has also indicated that the availability, access and physical/social attributes of such aesthetics and safety of spaces for active leisure are influenced by a neighbourhood's socio-economic level and that low SES areas possess fewer facilities of poorer quality than the average (Vieira *et al.*, 2013). Such a thesis aligns well with Sibley's (1995) 'geographies of exclusion'.

The term 'geographies of exclusion' was coined by Sibley (1995) and openly influenced by the works of David Harvey, whose work in turn draws heavily on Marx and Engels and their theories about capitalist society and capital accumulation. Engels famously said that the human landscape can be read as a landscape of exclusion. Marx illustrated in *Capital* how capitalism uses space to solve its internal contradictions and Harvey further developed this argument to explain various ways in which, in modern capitalist societies, socio-spatial issues reinforce exclusion. Two examples offered by Harvey and Sibley are shopping malls and condominiums. Their focus has concentrated more on spaces that are purposely designed to exclude, albeit in a veiled form. In general though, less attention has been paid to the development of spaces within poor communities that reinforce marginality and exclusion. Such spaces have been investigated before, social housing being an important case in point (Hackworth, 2008). However, to date, little attention has been paid to urban transformations as enacted through the staging of mega-events. In this case, the geographies of exclusion may apply to communities that have not been directly displaced or gentrified but that nonetheless have suffered severely from associated urban 'renewal' projects.

In an effort to address this concern, a series of studies was conducted in low SES communities in Rio de Janeiro before, during and immediately after the 2016 Olympic Games. These studies focused particularly on exploring the impacts of the urban transformations taking place in the city for the hosting of the 2016 Rio Olympic Games on leisure spaces available to the community, and on residents' engagement with these spaces (Reis *et al.*, 2013, 2014, 2016; Sousa Mast *et al.*, 2013, 2016, 2017). We found that leisure spaces within these communities reflect the broader exclusion patterns faced by low SES communities in Rio de Janeiro and other major urban centres around the world. They are of poorer quality, access is significantly impeded by violence, they lack maintenance, are degraded and are frequently ephemeral as disjointed infrastructural and social projects come and go, accompanying any political change. Mega-events are one of the various influences that these communities experience in the midst of discourses of 'regeneration' of disadvantaged communities, but are nonetheless a very significant one.

(Continued)

Case Study 5.1: (Continued)

These studies challenge the future of mega sports events in two ways. First, if we understand leisure as being dependent upon opportunities that are created in advance of the experience (Elias & Dunning, 2008), and that leisure is central to the very idea of mega sport events, the spaces in which this social practice can take place are critically important to urban transformation projects. Second, the democratisation of access to leisure spaces, including the permanency of physical 'realities' and the consequent creation of meaningful ties, must be taken seriously by event organisers and governments that support their hosting. The experiences reported so far suggest a significant disregard for those living on the margins, being frequently 'moved on' (Lea *et al.*, 2012) by urban renewal projects through displacement or gentrification. As Sibley (1995: ix) argues, 'power is expressed in the monopolisation of space and the relegation of weaker groups in society to less desirable environments'. By carelessly modifying the urban environment to create a stage for hosting athletes, officials and tourists, mega-event organisers and the political institutions behind them are allowing the perpetuation of exclusion that is not only geographically located but also geographically reinforced.

Key Reading

Sibley, D. (1995) *Geographies of Exclusion: Society and Difference in the West*. London: Routledge.

Sousa Mast, F.R., Reis, A.C., Vieira, M.C., Sperandei, S., Gurgel, L. and Pühse, U. (2017) Does being an Olympic city help improve recreational resources? Examining the quality of physical activity resources in a low-income neighborhood of Rio de Janeiro. *International Journal of Public Health* 62 (2), 263–268.

Additional references cited in this case study are included in the reference list.

The concept of a sport centre gives rise to the concept of a sport tourism centre, sport tourism destination or sports resort. The attractiveness of sport tourism centres may draw upon the uniqueness of different sports regions that exist within a country (Rooney & Pillsbury, 1992). By definition, a sport tourism centre requires the presence of sports facilities and resources as well as tourism infrastructure and services (Standeven & De Knop, 1999). 'To the visitor the amenities appear to be related to each other; the whole is more attractive than each separate amenity' (Dietvorst, 1995: 165). Sport tourism centres have the capacity to accommodate significant inward travel flows at a destination. An established tourism economy in the form of national and/or international transport nodes, an established accommodation sector, tourist attractions to complement the sport industry and a well-developed service sector including tourism information services is critical to its functionality (Whitson, 2004).

The development and management of sport tourism centres and regions require the strategic and coordinated development of both sport and tourism resources (Maier & Weber, 1993; Pigeassou, 2002), with careful consideration given to strategic development opportunities, which may enhance the status of sport tourism centres. International Football Camp Styria is an excellent example of strategic sport tourism development. Located in the Alps region of Steiemark (Austria), Camp Styria offers integrated high-altitude football training venues that target elite football clubs and international football teams to prepare for pre-season league (summer camps) and international competition. In this case, altitude and world-class training facilities provide a competitive setting for the physical conditioning and preparation of sports teams, with Graz serving as a convenient regional air transport gateway. Camp Styria has become a sport centre, offering a combination of sport and tourism resources to serve a specialist elite sport market.

The evolution of sport tourism in central locations

Spatial change within the sports industry takes place continually within the urban milieu (Bale, 1989). This is particularly evident in the locational dynamics of team sports in Europe and North America. Sports stadiums in Britain were originally located to take advantage of population concentrations and transport nodes. The strategy of minimum aggregate travel for sport spectators resulted in the development of sport stadiums in inner-city locations. Urban hub-and-spoke public transport networks brought the majority of supporters relatively short distances by train and bus to attend sports matches at central locations. However, these locational criteria have lost much of their relevance given increasing stadium size and the need for safe, seated stadiums following the introduction of strict stadium safety regulation in the 1980s (Bale, 1993; Paramio et al., 2008).

The situation of locational flux arrived relatively belatedly in British sports, although in recent times crowd safety and revenue generation have driven the development of new stadiums, facility sharing and relocation in various sports (see Focus Point 5.2). By contrast, the situation in North America 'has been characterised since 1950 by a state of locational flux' (Bale, 1993: 150). Over the past 50 years, many high profile professional football, baseball and ice hockey franchises in North America have been relocated from one city, state or country to the next. Here, a different scale of analysis applies as entire teams are relocated between cities in different states and countries. The geographic delineations of sport competition have changed dramatically through this process. The expansion of professional ice hockey franchises to the warm weather climates of California and Florida, and the relocation of teams between Canada and the United States, aptly demonstrate this point.

Focus Point 5.2: Sports Stadiums and Locational Flux

Deliberations relating to the location and design of a stadium have important implications for a sports team in terms of player recruitment, the interests of local fans, heritage and the extent to which a sports team may function as a tourist attraction. In the English Premier League (EPL) names like Highbury, Maine Road and Upton Park (Arsenal, Manchester City and West Ham United, respectively) were iconic venues steeped in sports heritage. However, these historic venues have been consigned to residential redevelopment in recent years, and the clubs relocated. Emirates Stadium became the new home of Arsenal FC in July 2006. Manchester City FC relocated to the Etihad Stadium following the 2002 Manchester Commonwealth Games and home to West Ham United FC is now London Stadium, which was constructed for the 2012 London Olympic Games. This cements a trend towards locational flux, long evident in North American professional sports, but a relatively belated development in the UK. Other London sport clubs have relocated completely in search of large, safe sport facilities. 'For more than 100 years Wimbledon FC called south London home. It was there, like any home, that foundations were laid, memories were created and a legacy was built... The club was forced to leave Plough Lane in 1991 – their home since 1912 – after the Taylor Report recommended that all top-flight teams should have all-seater stadiums' (*The Guardian*, 2016: online). After a long and proud history at Plough Lane (including the FA Cup final victory over the mighty Liverpool FC, the dominant English Premier League club of the 1980s, at Wembley Stadium on the 14 May 1987), and a decade of ground sharing with Crystal Palace (Selhurst Park, London), Wimbledon FC was relocated to Milton Keynes (56 miles to the north-west of London) and renamed the Milton Keynes Dons. The loss of heritage associated with this move and name change resulted in the complete dislocation of lifelong club fans, leading supporters to form a new club (AFC Wimbledon) at the original venue (Plough Lane, Wimbledon). Locational flux has now extended to other sports in the UK. The Wasps Rugby Club announced in October 2014 that from December of that year it would be hosting its 'home' games at Ricoh Arena in Coventry, a multipurpose stadium and home to Coventry FC, that was developed in association with conference, training, banqueting, exhibition, music and sports facilities, a casino and shopping centre, Tesco supermarket and hotel accommodation. In terms of market range and accessibility, it is claimed that Ricoh Arena, Coventry, is centrally located within a two-hour drive of 75% of the population of England. This is, however, for local fans 81.3 miles (130 km) via the M40 from the club's former venue in High Wycombe (north-west of London) and, as the case of Wimbledon FC amply demonstrates, the loss of heritage associated with the relocation of a sports club should not be underestimated.

Sources

The Guardian (2016) The Resurrection of AFC Wimbledon. The Guardian online (18 May 2016). See https://www.theguardian.com/football/copa90/2016/may/18/afc-wimbledon-players-fans-manager-league-two-play-off (accessed 8 September 2017).

Ricoh Arena. See http://www.ricoharena.com/about-us/ and https://en.wikipedia.org/wiki/Ricoh_Arena#Before_Wasps.27_Relocation (accessed 8 September 2017).

The heritage values associated with sports teams are invariably compromised in the process of inter-city or transnational relocation (see Chapter 6) and may, in fact, be lost to a region altogether if transferred to a new host city (Kulczycki & Hyatt, 2005). Ironically, the situation of locational flux has in many cases taken place in association with the development of the tourism product, including tourist attractions that target the nostalgia sport tourist (Ramshaw & Hinch, 2006; Stevens, 2001) (see also Chapter 10). Nostalgia sport tourism has been actively developed in North America (Rooney, 1992). Sports halls of fame have often been positioned as tourist attractions with traditional museum-style presentations being succeeded by new generation sport attractions, which feature cutting-edge interpretive techniques, designs and technologies. This has not been possible given the static location of facilities in other parts of the world. Stevens (2001) notes, in reference to sport halls of fame in England, that their

> locations tend to be governed by non-market related criteria, such as location of administrative offices or the owners desire to convert a hobby into a public display. Most are located outside the major metropolitan areas, and when compared to the geography of major league franchises, and hence major stadium developments, it is apparent that the opportunity to physically link sports stadia with visitor attractions has largely been missed. (Stevens, 2001: 69)

The Spatial Analysis of Sport Tourism in Peripheral Locations

Christaller (1963/64: 95) states that tourism is 'a branch of the economy that avoids central places and the agglomerations of industry. Tourism is drawn to the periphery ... (where) one may find, easier than anywhere, the chance of recreation and sport'. Sport tourism in peripheral locations (Chapter 7) is typically focused on natural resources. Examples of these resources include mountains, lakes and rivers that form the resource base for sports such as mountain climbing, skiing, rafting, kayaking and angling (Gilbert & Hudson, 2000). Sport tourism in peripheral locations is typically resource dependent and thus determined by the physical nature of the landscape rather than proximity to market areas. Sport tourism market zones, travel patterns and tourist experiences in peripheral locations stand in contrast to those associated with sports that take place in central locations. The principles governing the spatial dynamics of sport tourism in peripheral areas are proposed in Table 5.2.

Sports space theory applied to peripheral areas suggests that the natural resource base, rather than market access will determine the locations where sport tourism takes place. A ski resort, for example, is dependent on the requisite elevation, terrain and snow conditions among other things, to allow participants to engage in their sport in favourable conditions (Hopkins, 2014). This is especially the case for niche sport tourism markets where specific sport motivations requiring unique natural environmental attributes often apply. As Bourdeau *et al.* (2002: 23) observe, 'the location of sites and itineraries thus depend on diverse natural conditions which do

Table 5.2 Spatial dynamics of sport tourism in peripheral areas

(1) The main challenge of the sports areas in the periphery is to facilitate visitor access and opportunities to engage in sports in natural areas. Sports areas are located in peripheral areas where natural resources and built infrastructures rather than centrality determine site location decisions.
(2) Peripheral sports areas are reliant primarily on active sport tourists as participants rather than spectators.
(3) The quality of the sport environment/resource, rather than the number of sports provided, determines the order of the peripheral area within a sports location hierarchy. Quality may be determined by uniqueness, naturalness/absence of impact, remoteness and features of the natural environment.
(4) Peripheral sports locations exist in clusters of critical mass allowing the development of a high standard and range of sports facilities that enhance the standing of the destination in the sports location hierarchy.
(5) Higher-order locations are clustered in peripheral areas where natural features and developed infrastructure and services facilitate sport tourism.
(6) Consumers of sport tourism in peripheral areas may be motivated by the desire to (a) engage intensively in their chosen sport and/or (b) maximise the other tourism opportunities associated with the pursuit of their chosen sport.

not readily lend themselves to the satisfaction of geographic (accessibility), demographic or economic needs'. The resource requirements of sports may be moderated through, for example, snow-making technology in the case of alpine winter sports. Resources such as artificial ski slopes can be constructed at considerable expense in central locations, with immediate access provided for concentrations of population. Notwithstanding these points, the resource requirements of sport tourism in peripheral areas remain the fundamental characteristic of the locations in which they take place (Higham & Hinch, 2009).

The inescapable circumstances of sport tourism in peripheral areas provide unique challenges in terms of commercial development (Bourdeau et al., 2002). Remoteness and terrain may limit access while reliance on weather conditions and climatic uncertainty may compromise the viability of sports or render them impossible in certain conditions. The consequences include seasonal use variations, low-intensity use due to institutional factors, high mobility of visitors between sites and self-sufficiency on the part of many users in terms of service requirements (Bourdeau et al., 2002). Where favourable natural resources and market access coexist, a competitive advantage may be achieved. The development of a critical mass of sport tourism activities and facilities in peripheral areas may stimulate further investment in transportation and infrastructures thereby improving access.

This discussion suggests that, like central locations, a hierarchy of peripheral sport tourism destinations exists (Table 5.2). Higher-order destinations are generally located in peripheral areas where natural features and developed infrastructure and services are present. Higher-order places may cement this desirable status by fostering and building unique cultures (and subcultural values) associated with specific sports. 'Depending on the resources that they offer, and their reputation and use characteristics, sites generally become established in a very clear hierarchy, in which they are identified as being of local, regional or national interest' (Bourdeau et al., 2002: 24). In their study of 2000 climbing sites in France, Bourdeau et al. (2002) identified a hierarchy in which 85% were considered to be of local, 13% of regional and 2% of national significance. However, the existence, and the functioning of a peripheral sport tourism destination, is also determined by the level of tourism infrastructure and services. In reference to the 1994 Lillehammer Winter Olympics, Teigland (1999: 308) states that 'the influence zone of a particular Olympic Games will vary depending on the distribution of venues in different types of satellite areas... (and) entry and departure points to the host country or region, especially areas close to airports receiving international visitors'. These principles of sport space assist sport tourism managers to understand the opportunities and potential, as well as the limitations that apply to active and event sport tourism development in peripheral areas.

The hierachy of peripheral sport tourism locations is fascinating to contemplate given the critical importance of natural resources, and the changing distribution of natural resources under climate change scenarios (Higham & Hinch, 2009). Hopkins *et al.* (2013) adopt the concept of relative vulnerability to frame their study of how climate change is likely to affect the perceived attractiveness of regional ski resorts in southern New Zealand, relative to their competitors elsewhere in New Zealand and in Australia. They find that although southern New Zealand ski resorts face the threat of a warming climate, with relatively favourable natural and artificial (snow-making) conditions, they are less vulnerable than North Island (New Zealand) and Australian ski areas. Their research highlights the '... clear need to move beyond a focus on snow reliability to consider the broad range of factors that contribute to regional variations in vulnerability (and) the critical importance of situating relative vulnerability within a social context' (Hopkins *et al.*, 2013: 449). There is no question that climate will have an increasingly heavy influence over not only the standing of destinations in the sports location hierarchy, but also the very survival of ski resorts (Hopkins, 2014; Scott & McBoyle, 2007; see also Chapters 7 and 9).

Conclusion

This chapter highlights the factors that shape and influence sports locations and the spatial dimensions of sport tourism travel flows. It extends the concepts of the sport centre and the hierarchy of sport locations (Bale, 1993) within the context of sport tourism. The prominence and status of sport tourism centres are determined by the range and quality of available sports experiences, in combination with levels of service development and unique sport facilities and resources. Sports event experiences can be usefully considered at various spatial scales from the global to the local (Pettersson & Getz, 2009).

Sports locations may command regional, national or international travel flows that can be actively influenced by sport and tourism organisations. The location of sport tourism activities in central or peripheral areas exerts a major influence on the market range, spatial travel flows and visitor patterns. An appreciation of travel flows that exist within the spatial dimension of sport tourism is fundamental to sport tourism development. The status of sports locations can be actively fostered in an attempt to develop new markets and extend market range (Higham & Hinch, 2002a, 2002b) through the implementation of development strategies. The redevelopment of Melbourne Park at the cost of $A383 million is a clear strategy to further enhance the global status of Melbourne as an urban sports location, as well as to retain the Australian tennis grand slam (the 'Grand Slam of the Asia-Pacific'), extend market range and expand visitor markets (Hede & Kellett, 2010). Key geographical theories and concepts invite critical consideration of the spatial dynamics and interrelationships that define and shape sport tourism phenomena (Higham & Hinch, 2006).

6 Place, Sport and Culture

> *We have essentially treated sport and tourism as cultural experiences—sport as a cultural experience of physical activity; tourism as a cultural experience of place. It will come as no surprise, therefore, that the nature of sport tourism,... is about an experience of physical activity tied to an experience of place.*
>
> Standeven and De Knop, 1999: 58

Introduction

Sport exerts a significant influence on the meanings that people attach to space. These meanings are central to the experience of sport tourists, to the impacts that they have on the destinations they visit and to the strategies designed to shape development. This chapter explores these claims by considering the unique nature of sport tourism places, examining sport culture in relation to place identity, and by addressing the way that sport is used to brand tourism places. Daniel Evans' case study (Case Study 6.1) considers the tensions around local football (soccer) identities in Liverpool and the global context of the game, illustrating the complex realities of place, culture and marketing in a contemporary sport tourism context.

Place

Place refers to meaningful space (Lewicka, 2011) whereas space is a landscape of geometry as manifest in location, area or distance. Crouch (2000) expands on the difference between space and place in a leisure context with his statement that:

> Space can be a background, a context, a 'given' objective component of leisure and tourism. In that way it is seen as a location, a National Park or a site where particular leisure/tourism happens, a distance between things. Place can be a physical image that can be rendered metaphorical as the content of brochures, 'landscape' as a foil for what people might imagine they do... In this way it may be that place is understood to be a cultural text that people read and recognize directed by the particular intentions of a producer or promoter. (Crouch, 2000: 64)

While the geometric characteristics of space can be objectively measured, place is much more subjective. Individuals and groups are constantly defining and refining the meanings that they attach to spaces. As other aspects of their lives change, so do the meanings that they attach to spaces.

Sport (Bale, 2002) and leisure (e.g. Lee *et al*., 2012) are two phenomena that infuse space with meaning. Just as Williams and Kaltenborn (1999: 215) have argued that outdoor leisure activities are undertaken 'to establish identity, give meaning to their (the participants) lives, and connect to place', we argue that sporting engagement functions in the same way. In one of the first texts in this field, Standeven and De Knop (1999: 58) contended that sport tourism is 'an experience of physical activity tied to an experience of place'. More recently, Gammon (2015) renewed this claim and has called for more research into the way that place is experienced in a sport tourism context. Examples of relevant research responses include Brown *et al*.'s (2016) study of sport involvement, place attachment, event satisfaction and spectator intentions at the London Olympics, in which they concluded that the psychological connection that spectators form in relation to event venues does not necessarily transfer to their attachment to the host city. Kirkup and Sutherland (2017) make a similar claim in their research, noting that the host cities of sport events are often challenged to attract repeat visitors beyond those returning for a similar event.

Sport tourists become attached to place in a number of different ways. While place attachment is multidimensional and complex, two of its most prominent aspects include place dependence and place identity. Place dependence is the functional tie that individuals and collectives have to a space. Brown and Raymond (2007: 90) state that '[p]lace dependence refers to connections based specifically on activities that take place in the setting'. In the context of sport tourism, a particular location may have a unique combination of resources that facilitate certain sporting activities. An example would be the dependence of downhill skiers and snowboarders on accessible ski resort areas that receive generous amounts of snow cover. This functional dependence typically contributes to a strong attachment to the area (see discussion on sport tourism in peripheral areas in Chapter 5). The second major dimension of place attachment is place identity, which plays a role in the self-making of individuals and groups (Kerstetter & Bricker, 2009; Scannell & Gifford, 2010). In articulating where we are or where we play, we add to our understanding of who we are. Traveling for sport is therefore an important part of the way we construct our self-identity. For example, ultramarathon runners are attached to the sites where they compete both as a result of their dependence on them to host the event (Hinch & Holt, 2017) and in terms of the importance of the destination as part of their place identity (Hinch & Kono, 2017).

Tourism places

In his seminal work, *Place and Placelessness*, Relph (1976) argued that the concept of sense of place was most applicable in the local environment where individuals are in a position to develop deep attachments associated with home. He suggested that tourists were one of the least likely groups to develop a 'sense of place' in relation to the destinations that they visit because of the superficial nature of their experience and the tendency of the tourism industry to present 'disneyfied' landscapes devoid of deeper meaning. In the 40 years since Relph's claim, the very concept of home has been challenged by increased mobility, which has contributed to new understandings of what 'home' is, including the possibility of having multiple homes. However, even in the traditional sense of home as implied by Relph, it has become a well-established fact that visitors to recreation and sporting locations not only 'can' develop strong connections to the places where they play, but often 'do' (e.g. Wynveen *et al.*, 2012). A focus on home-based connections to place also ignores the importance of the 'social worlds' that help to bond visitors to leisure-oriented destinations (Kyle & Chick, 2007). Similarly, Stebbins's (2007) theory of serious leisure suggests that such commitment is often characterised by travel careers that connect individuals to sport tourism destinations (Jones & Green, 2005). Finally, Tuan (1975: 165), another pioneer in place attachment, argued that to develop attachment you must live in a space as '[t]o live in a place is to experience it, to be aware of it in the bones as well as with the head'. Given the embodied nature of sport-based travel (e.g. Lamont, 2014), it can be argued that by engaging in sport during one's travels, one can, in fact, literally feel these places in one's bones thereby connecting travellers to place.

Travelling for pleasure beyond the boundaries of one's routine life-space implies that there is some experience available at the destination that cannot be found at home, and which compensates for the costs of the trip. Standeven and De Knop (1999) presented this line of argument by suggesting that the:

> nature of tourism is rooted in authentic cultural experience of places away from home that have different characteristics. Those characteristics are unique to each place, and the tourist views, feels, hears, smells, and touches them. Their differences (and their similarities) become a part of his or her conscious experience. (Standeven & De Knop, 1999: 57)

The destinations where authentic experience occurs become infused with meaning for those who visit them. Gu and Ryan (2008) likewise argue that tourism development can impact the locals' place identity based on their perception of distinctiveness, self-efficacy, continuity and self-esteem associated with the place. All of these dimensions can be impacted by the type of tourist activity occurring at that place. Such logic suggests that place is integral to sport tourism. From the perspective of the tourism industry, the more meaningful that a destination is for visitors in a positive sense, the greater that destination's competitive advantage in the tourism market place. Sport tourism destinations that foster attachment through place identity as well as place dependence are less likely to be substituted for competing destinations in the future (Hinch & Holt, 2017).

Sport tourism places

The experiences of sport tourists '– staged or real – result from tourists' interactions with place' (Standeven & De Knop, 1999: 58). Bale (1993) described four ways that these interactions with place occur in a sport context: (1) a search for the sacred, (2) the development of home-like ties to a destination, (3) aesthetics and (4) sport heritage. In the first instance, Bale (1993) argues that there has been a changing 'religious' allegiance for a substantial portion of the public. Rather than worshipping at a religion's alter, many people have substituted sport as their alter. Like religious pilgrims, sport tourists travel from the profane (origin) to the sacred (sport destination) and back to the profane (origin) (Graburn, 1989). Sport tourism sites act as a refuge from modernity and the sacred forms a reality separate from the ordinary lives of travellers, perhaps reflecting a response to conditions of rootlessness in a post-modern world.

A second way that sport spaces become endowed with meaning is through the development of home-like ties to a site even though that site may be far removed from one's residence. A particular sporting venue may become home as fans or active participants develop sporting allegiances to the site. This idea of 'home' contrasts with most technical definitions of the tourist, which provide an arbitrary distance threshold that once surpassed, define a traveller as a tourist or excursionist. In doing so, interesting questions are raised about the meaning that these 'home' fans attach to the destination (Case Study 6.1).

Case Study 6.1: Local Identities in a Global Game – Liverpool's Football Space

Daniel Evans, York University

The restructuring of English football and its commercialisation on an unprecedented global scale have encouraged particular cities – especially Liverpool, UK – to increasingly orient local economic development around football. The changing nature of consumption has made television revenues a key priority (Deloitte, 2014; Gibson, 2012). But live consumption remains an essential component in the cultural production of the English Premier League (EPL) experience by engaging audiences in the practices that produce these fabled spaces of football.

(Continued)

Case Study 6.1: (Continued)

Overseas supporters benefit teams through merchandise sales, by increasing television ratings and sometimes by making the pilgrimage as football-tourists to watch the team in person and take part in the rituals they have witnessed from afar (Edensor & Millington, 2008). Though their visits are of short-term duration, expenditures are relatively high and, therefore, valuable to the local club and host city. Meanwhile, local fans remain enthusiastic supporters because they are emotionally invested in a club that has come to be part of their own identity. Both local and overseas fans demonstrate strong place attachment through their fandom.

Local supporters both love and loathe other fans of their team. On the one hand, a large and passionate fan base can affect results on the field and provide a team with the financial resources to secure the talent necessary to compete at the highest level of the game. On the other hand, tourists are not very active in representing the club spaces and are unfamiliar with many of the symbols and songs. Everton FC's fans, the cross-town EPL rivals to Liverpool FC, pride themselves on their 'localness' and mock Liverpool FC's larger contingent of tourist fans, although they are quite willing to accept non-locals into their own fold. But growing a legion of foreign fans directly threatens regular supporters' access to matches (Kerr & Emery, 2011). Access for local fans is important for the participatory aspects of fandom; through noise and visual display, local fans have an affective influence on other fans and the club. Therefore, a member of 'LFC Supporters in Toronto' has different expectations and expressions of support for the club than a member of 'Keep Flags Scouse', even if they share a common interest in supporting Liverpool FC. Self-awareness of their own agency is encouraging some supporter groups to increasingly organise and regulate their own actions. Informal surveillance and self-policing in the stands encourage particular modes of behaviour by supporters in different sections of the stands (Dixon, 2014).

The practices of fans take on a particular spatial structure. How fans shape and occupy football space is not the only dimension of how that space is socially produced. In Liverpool, this organising of football space is becoming critical because the sport is no longer seen as just an amusing pastime, but a key plank of the city's economic base; it is a city building itself for football. The importance of football then creates its own space where, in Lefebvre's (1991) framework, football is perceived, conceived and represented as vital to the city.

Liverpool is building a football space created and shaped by the clubs and their many supporters. Fans are able to participate in the cultural event by attending a match or forming part of the audience who watch the televised event. Clubs reinforce their brand both through the active use of their marketing as part of the event and through the less direct use of branded fans and supporters, who also form a part of the clubs' revenue streams through the merchandising of various paraphernalia (Kerr & Emery, 2011). In Liverpool, it is clear that this is a dynamic process with constant mediation between the groups of fans and the clubs. But there is a distinct spatial element to this negotiation of club identity. It embraces what is considered home, what territory belongs to the club and its adherents and how a network of fans and supporters is diffused through the rest of the football world. Groups in Liverpool – and elsewhere – have sought to intervene in the cultural production of football in their city by giving their sport a glocalised version of the product, one that is recognised worldwide for its particular performance of fandom. The local fans of both clubs have to negotiate how it is they deal with an influx of global fans. The local cannot be separate from the global, since it is the local identity of clubs that fascinates foreign audiences (Salazar, 2005). Would Liverpool FC or Everton FC be as popular if they were exactly the same as all the other teams? No. It is the specific football space created in Merseyside that has contributed to the global following of those teams. The football space of Liverpool is a social and cultural space that extends beyond its economic value. The contestation of this space by locals is an attempt to control the production of football space and assert that the global game can have a Liverpool-specific and even a club-specific inflection that is, in their view, best managed by the hands of the locals that created it.

Key Reading

Edensor, T. and Millington, S. (2008) 'This is our city': Branding football and local embeddedness. *Global Networks: A Journal of Transnational Affairs* 8 (2), 172–193.

Gibson, O. (2012) Premier League Lands £3bn TV Rights Bonanza from Sky and BT. *The Guardian*. See http://www.theguardian.com/media/2012/jun/13/premier-league-tv-rights-3-billion-sky-bt.

Kerr, A.K. and Emery, P.R. (2011) Foreign fandom and the Liverpool FC: A cyber-mediated romance. *Soccer and Society* 12 (6), 880–896.
   Additional references cited in this case study are included in the reference list.

A third way that sport spaces may become endowed with meaning is through aesthetics. In this case, place meaning is derived from a variety of sporting landscape elements that contribute to the aesthetics of a sport place. For example, football stadiums in the UK have been described as 'secular cathedrals' that are intimately tied to perceptions of place, both from the perspective of the followers and non-followers of the sport (Robinson, 2010). More tangibly, temporary 'live sites' and 'fan parks' in association with major events (McGillivray & Frew, 2015) or the development of permanent sports city zones or quarters represent a manifestation of the aesthetics of sport on the tourism landscape. Smith (2010) suggests that permanent urban sport zones need to go beyond a focus on major event facilities to include other more participatory sports facilities such as halls of fame, exhibitions of sport activities and facilities of active engagement.

The last element in this typology is sport heritage (Chapter 10). The concepts of sport heritage places and tourists fit well with the idea of nostalgia sport tourism (Fairley & Gammon, 2005; Gammon & Ramshaw, 2007). Support for this view is found in the prevalence of sport museums, tours of former Olympic sites and visits to the origins of various sports such as golf at St Andrews, Scotland. Often, the emphasis on this element is the built environment, especially the bricks and mortar of historic sports facilities that serve as tangible icons of sporting pasts. More subtly, however, the cultural impacts of sport on the natural landscape can be a powerful dimension of place whether they are expressed in spaces of traditional sport practices such as purpose-built arenas or in areas designed for other uses but which are co-opted for sport such as residential streets used for informal sporting practices and pastimes.

While all of these factors infuse sport spaces with meaning, thereby creating sport tourism places, Bale (1989) notes the emergence of sportscapes (landscapes characterised by standardised sport facilities and infrastructures) as a counter trend. In fact, these sportscapes are an embodiment of 'placelessness' as described by Relph (1976).

> In the twentieth century sportscapes rather than landscapes have tended to characterize the sports environment... New materials had changed the shape of the stadium and the texture of the surfaces; fields became carpets and parks became concrete bowls. Most sports require artificial settings, although the degree to which the natural environment needs to change varies between sports. (Bale, 1989: 145)

The rationale for this trend towards standardised sportscapes is at least fourfold. First, it represents an attempt to ensure spectator and participant comfort and safety; a powerful motivation in communities that increasingly turn to the courts to establish liability and financial penalty for personal injury at sporting sites. Secondly, the standardisation of sport facilities and sites is seen to provide an 'even playing field' and in so doing, to foster fairer competition. Thirdly, the evolution of sportscapes is a reflection of technological advances that have allowed for 'improved' performance and therefore tend to be mimicked in competing regions. Finally, the trend towards sportscapes is often an outcome of mass media broadcast requirements both in terms of technological needs and in terms of market appeal. Notwithstanding this trend towards conformity, there has been a conscious effort over the past decade to design major sport stadiums in ways that tie them meaningfully to the locations in which they exist (Sheard, 2014; also Case Study 4.1).

Territoriality

The issue of territoriality arises when two or more groups who feel that they 'own' or have some sort of moral claim over a site, come into conflict. Significant differences in the way these groups identify with a place, the pressures of crowding and feelings of loss or invasion have contributed to the growing challenge of territoriality as a dimension of place that functions in opposition to sport tourism. This issue has been studied in the context of spectators at sport events (Evans & Norcliffe, 2016; Case Study 6); residents and visiting athletes at participation-based events (Hinch & Holt, 2017); and in terms of different types of subculture members in sports such as surfing (Usher & Gomez, 2016). While not referring to territoriality directly, Evans and Norcliffe (2016) highlighted the tensions between the local identities and globalisation in soccer with their study of the strains between visiting and local supporters of Liverpool Football Club (also Case Study 6.1). Despite the fact that these groups were cheering for the same team, local supporters were conflicted in terms of their identities and relationships with visiting supporters. Hinch and Holt's (2017) study of the place attachment of residents and visiting athletes at a participant-based ultramarathon event found that both groups felt that the event had a positive impact on their attachment to the host community. However, the athletes were more likely to be attached through place dependence while the locals were more likely to be attached through place identity. In a quite different sporting context, Usher and Gomez's (2016) study of surf localism in Costa Rica identified three types of surfers including locals, foreign resident surfers and short-term visiting surfers. Native Costa Rican surfers felt a greater sense of ownership of the surf but were less likely to openly challenge short-term visiting surfers who violated surfing etiquette than were resident foreigners. These conflicts highlight the need to manage the interactions between various surfing groups so that conflict is minimised.

Culture, Place and Identity

Place is closely tied to culture. The meanings that are attached to sport spaces are strongly influenced by the cultural context in which sport and tourism exist (Funk & Braun, 2006). Culture relates to sport in a number of ways, but three of the most tangible associations include: (1) cultural programmes run in association with sport events, (2) sport as a form of popular culture and (3) subcultures in sport. Each of these cultural dimensions influences the meaning that is attached to sport spaces and in so doing, they affect place identity and potentially, place making for tourism.

Sport and culture

From 1912 to 1948, the Summer Olympic Games featured a 'pentathlon of muses' in which artists competed in contests related to architecture, musical composition, sculpture, painting and literature (Ingraham, 2016). Despite the demise of these competitions, the 'cultural arts' remain a fixture at the Olympics particularly as manifest in the Opening and Closing Ceremonies and the associated arts programmes. This practice reflects a popular approach of treating sport and culture as separate but complementary activities particularly at mega sport events.

The three types of narrative approaches associated with the opening ceremonies of major sporting events include: history, party and show (Moragas Spa et al., 1995). In the first case, the ceremony is treated as a 'unique historic event taking place in that moment, although forming part of a historic chain' (Moragas Spa et al., 1995: 105). The opening ceremony of the 2012 London Summer Olympics was a good example featuring a mosaic of historic elements that created a narrative of national identity (Baker, 2015). In the second instance, the ceremony is treated as a celebration of the performing arts that highlight the local, regional and national performing arts scene. It is a joyous explosion of the local arts in celebration of the event. Finally, the third type of ceremony is one of entertainment. This type of ceremony downplays the 'distraction' of the cultural and ritual structures of the event and tries to provide 'an entertaining introduction to the "real" excitement: the sports competition' (Moragas Spa et al., 1995: 108). All three of these approaches treat sport as separate from culture.

The 'fine arts' programme held in conjunction with the sporting competitions of the Olympic Games is another example of the distinct but complementary association of sport and culture (García, 2010). Article 39 of the current Olympic Charter (IOC, 2015) requires that a cultural programme be held under the auspices of the local host committee although its application has varied considerably from games to games. Despite the ambitious cultural programmes that have been proposed at the bid stage, the delivery of these programmes has been much more modest. As such, the cultural component appears to be more important as part of the sales pitch to win the bid than it is in the actual delivery of the event. If these cultural components are truly valued, this needs to be articulated more clearly by the IOC and similar bodies. More financial support should be provided and the host city should be held accountable for their commitment to this programme (García, 2010).

Sport as culture

Popular culture as manifest in sport is one of the main ways that humans develop personal and collective identities (L'Etang, 2006). It is through these personal and collective identities that place identity is developed. At its most basic, identity is the way that we perceive ourselves, as individuals and collectives, based on prevailing social and ideological values and practices (McConnell & Edwards, 2000). It is a social phenomenon developed through social and cultural processes as found in traditional media, social media and other dominant cultural institutions. Identity is the way in which people make sense of the self through affiliation and bonds with other people and the cultures that define these affiliations (Dauncey & Hare, 2000).

National identity is typically thought of in the way that nations differ from each other in terms of stereotypes, symbols and practices including those associated with sport (Devine & Devine, 2004; McConnell & Edwards, 2000; Tuck, 2003). McGuirk and Rowe (2001) articulated the way that place becomes part of our national identities:

> Places have come to be conceptualized as constructed through a dynamic articulation of their material and representational dimensions, and place identity is understood to be mutable, contingent and fluid. Cultural stocks of knowledge about places can, however, constitute a prevailing, often stubbornly persistent balance of forces that name, interpret and project place meanings. A place in this sense is a 'text', the meaning of which is continually being made, reproduced and re-made. (McGuirk & Rowe, 2001: 52, 53)

Place identity is influenced by many cultural attributes but sport certainly appears to be one of the most dominant (Porter & Smith, 2013). Nauright (1996) suggests that not only is sport a factor in the process of constructing place identity, but that it

> is one of the most significant shapers of collective or group identity in the contemporary world. In many cases, sporting events and people's reaction to them are the clearest public manifestations of culture and collective identities in a given society. (Nauright, 1996: 69)

It is not just high profile sporting competitions that influence and reflect place identity. Sports and leisure pursuits that occur on a daily basis in local communities are also important (Williams & Champ, 2015). Typically, insights into cross-cultural differences related to sport are gained incidentally as a by-product of travelling to another culture as a sport participant or spectator, or by being exposed to cultural differences through the media. Increasingly, however, sport is consciously being used as a lens to develop an understanding of cultural differences. This is exemplified by sport-based study-abroad trips by US universities (Fairley & Tyler, 2009).

Place identity through sport is constructed in at least four ways that have particular relevance to tourism. These include: (1) the association of particular sports to specific regions, (2) the unifying forces of competitive hierarchies found within sport, (3) the identification with sporting success and (4) the personification of place through sporting heroes and heroines.

Specific sports are commonly associated with particular nations. This connection may be based on a variety of factors, but one of the most powerful is the role that a given sport has played in a nation's heritage. An example of this type of association is that of rugby union in New Zealand. Falcous and Newman (2016) point out that such sporting identities can be based as much on myth as on reality.

The competitive hierarchy that exists in many sports is also an important factor in the promotion of place identity (McGuirk & Rowe, 2001). The principle reflected in the Bedouin saying 'I against my brother, I and my brother against my cousin, I and my brother and my cousin against the world', reflects the aggregation of territorial interest that occurs within a competitive hierarchy (Fougere, 1989: 116). Place identity is fostered through a growing territorial scale as successively higher levels of the competitive hierarchy are reached. In this process, many of the real differences and disparities that are found within these places are overshadowed or subsumed.

The relative success of a region in terms of its sport performance also influences the connection between sport and place identity. In addition to being in the 'news' more frequently, places with teams or athletes that consistently win major championships tend to be characterised as winners in their own right. Such success can provide a sense of common identity even when numerous other social and economic divisions may exist within the region (Delia, 2015; McGuirk & Rowe, 2001). One has only to look at the national celebrations associated with World Cup Football success to see evidence of this dynamic (Dauncey & Hare, 2000). Increased mobility and immigration has complicated the traditionally narrow focus of national identity in sport. Hinch et al.'s (2018) study of the national reaction to the success of Japan's ethnically diverse national rugby team highlights these issues (Focus Point 6.1).

Focus Point 6.1: Japanese Rugby and National Identity

Despite the processes of globalisation, sport remains one of the most prominent manifestations of national identity. The Olympics and the FIFA World Cup are prime examples of this as nations rally around their respective teams and athletes, sharing in the glory of their successes and the pain of their defeats. Rightly or wrongly, these athletes are often seen as embodying their home country's national character especially when they are enjoying success on the international stage. The Japanese national men's rugby team represents a dramatic example of the dynamic interplay of globalisation and patriot sport. With the 19 September 2015 surprise victory of Japan's rugby team over the favoured South Africans, the team surged in popularity on the home front. The continued strong play of the team and the forthcoming 2019 Rugby World Cup scheduled for Japan have ensured that the team has remained popular in Japan. Yet, even a casual observer can see that many of the names and faces on the team do not fit the typical image of Japanese players. The fact is that over one-third of the team were born outside of Japan although they all meet the criteria for inclusion on the national team (Hinch *et al.*, 2018). Like many other sports, rugby in Japan is a cultural resource that is used in the discourse of Japanese national identity. Sociological literature on Japanese identity suggests that the most predominant normalising discourse is the taken-for-granted assumption of an ethnically homogeneous and mono-cultural Japanese society (Befu, 2001; Burgess, 2010). It therefore seems paradoxical to see a heterogeneous mix of ethnic traits on the team while the team is promoted as a mechanism for Japanese national identity. The 'Brave Blossoms', as the team has been affectionately nicknamed, are challenging the traditional view of national identity and have become a focal point of identity politics in Japan. The way that this debate plays out will be one of the more fascinating backstories of the 2019 Rugby World Cup. The world's sporting eyes will be focused on Japan during the event as the complex issues of Japanese identity are negotiated on the international stage.

Key Reading

Hinch, T.D., Higham, J.E.S. and Doering, A. (2018) Sport, tourism and identity: Japan, rugby union and the transcultural maul. In C. Acton and D. Hassan (eds) *Sport and Contested Identities: Contemporary Issues and Debates* (pp. 191–206). London and New York: Routledge.

   Additional references cited in this case study are included in the reference list.

Finally, sport heroes and heroines can have a strong impact on the way that we identify with place (Dauncey & Hare, 2000; Nauright, 1996). A good illustration of this was the 2012 media frenzy coined 'Linsanity' that erupted when Jeremy Lin, the first Taiwanese-American to play in the National Basketball League, emerged from 'nowhere' to spark the struggling New York Knicks (Su, 2014). While the US media portrayed Lin paradoxically as an American underdog yet perpetual foreigner, the Taiwanese media portrayed him as a national hero who illustrated the global relevance of Taiwan. In both cases, Lin's star status served as a rallying point in the collective identities of basketball fans in the United States and Taiwan.

Sport subcultures and social worlds

Sport subcultures represent a third cultural dimension of sport that contributes to place identity in a tourism context (Focus Point 6.2). These subcultures are generally characterised by a commitment to a particular sport, distinguishing symbols or cultural capital and various career stages in terms of subculture membership. The use of the term 'subculture' in this instance is meant to be inclusive of lifestyle and neo-tribal concepts of sporting cultures that are distinguishable from dominant cultures (Wheaton, 2007). It is also closely related to the concept of social worlds (Unruh, 1980). Sport subcultures are of interest because they are characterised by unique relationships to place including place identity.

Focus Point 6.2: The Kosti Cycling Rally at Denpasar, Bali, March 2017

Glen Norcliffe, York University

The Komunitas Sepeda Tua Indonesia (KOSTI – Indonesia's veteran cycle club) invited members of the International Veteran Cycle Association (IVCA) – who are mostly based in Europe and North America – to attend their annual rally held in March 2017 at Kediri in East Java. Approximately 7000 cyclists attended, the great majority being Indonesians drawn from 700 KOSTI clubs representing 17 of Indonesia's 32 culturally diverse provinces, plus a few invited IVCA members. Most bicycles were Dutch machines imported during the colonial period or since then made in Indonesia. Attendees paraded along a 20 km route through the town, most in costume, many playing local and international music as they cycled along in club groups. This rally illustrates the meeting of numerous subcultures at a contemporary sports tourism event. Accommodation ranged from low-cost to expensive. Many attendees slept under trucks driven by local club members to Kediri, some chose multiple-occupancy rooms at a friend's house or a budget B&B using roll-out mattresses on the floor, and fewer lodged in hotels ranging from the budget to premium class. Food, likewise, varied from self-catering at low-cost to superior Indonesian restaurants. Subtle currents of post-colonialism and neocolonialism were in evidence. Currently, Indonesia has more veteran cycle enthusiasts than any other country which is partly a legacy of its status as a former Dutch colony. But economic colonialism was in evidence with some rally attendees from abroad staying in Western-owned hotels and consuming Western products at Western franchises. A collective enthusiasm for bicycles and cycling allowed the KOSTI rally to unite different classes in a common space. IVCA visitors were warmly welcomed and photographed with KOSTI groups numerous times. Indonesians attending represented a wide age range, but almost all grew up in the post-colonial era and enthusiastically welcomed the foreign visitors to their 'fellowship of the wheel'. However, the rally played a paradoxical role with respect to the cultural diversity of Indonesia. On the one hand, many KOSTI club members dressed in ways that presented their local culture, ranging from tribesmen in Papua to Sulawesi's traditional dancers. On the other hand, global influences present at the rally tended to undermine local cultures and diffuse homogenising cultural trends in ways anticipated by Pickel-Chevalier *et al.* (2016). Most foreign participants were retirees participating in a subculture, enjoying an active lifestyle by engaging in a sport that, like lawn bowling, cross-country skiing and pickleball, can be continued to an advanced age. Finally, the rally presented several environmental contradictions. The carbon imprint of 7000 people travelling from across Indonesia to Kediri was substantial, even though the bicycle is seen as an environmentally friendly machine. But cycling has played a significant role in establishing vehicle-free Sunday mornings on streets in Indonesia's major cities as environmentally friendly cultures have gained support.

Key Reading

Pickel-Chevalier, S., Violier, P. and Sari, N.P.S. (2016) Tourism and globalisation: Vectors of cultural homogenisation? (the case study of Bali). *Advances in Economics Business and Management Research* 19, 452–457. See http://www.atlantis-press.com/php/pub.php?publication=atf-16.

For example, the subculture of windsurfing has been described as a culture of 'conspicuous commitment' (Wheaton, 2000). This commitment is expressed in a number of ways but it is essentially reflected in the prowess, dedication and skill that members demonstrate in relation to their sport. Identifiable communities form around sports like mountain biking (Moularde & Weaver, 2016) and distance running (Allen Collinson & Hockey, 2007; Shipway *et al.*, 2012) with each community sharing basic characteristics that go beyond the sporting activity itself. Green and Chalip (1998: 280) highlight the relevance of these sport subcultures in their description of women's flag football as a sport subculture that '... gives participants much more than the opportunity to play together. It is a statement about who they are and the conventions by which they refuse to be constrained'. In cases such as surfing (Doering, 2018), snowboarding (Thorpe, 2011) and adventure sports in general (Breivik, 2010), sport subcultures represent a form of 'counter-culture' in that members deliberately distance themselves from the mainstream norms and practices of society. Subcultural sport activities give members the opportunity to develop their identities and to celebrate their sporting community (Chalip, 2006).

Beyond sporting engagement, membership in a subculture is normally reflected in a broad range of behaviours and practices. Wheaton (2000) articulates many of these with her statement about the windsurfing lifestyle group:

> Participation in this (subculture) lifestyle is displayed in a range of symbols such as clothes, speech, car, and associated leisure activities; however, for the dedicated, often-obsessive participant, windsurfing participation is a whole way of life in which windsurfers seek hedonism, freedom and self-expression. For 'core' members..., windsurfing dictates their leisure time, their work time, their choice of career, and where they live. (Wheaton, 2000: 256)

Shipway (2008) takes a trans-situational perspective in his typology of the distance running social world (Table 6.1). He describes sporting outsiders as not being directly involved in the subculture. As a result, they had little or no appreciation of the cultural meanings found within this group. Occasional or casual participants have a basic understanding of the sport but only engage in it infrequently. As such, they have a limited appreciation of the cultural meanings of the group. Regular participants value the sport but the sport does not dominate their identity. However, given their regular involvement, they have an appreciation of the cultural dimension of this group. Finally, insiders are experienced and committed members of the sporting subculture. They are therefore intimately familiar with the rules and rituals of their sporting community and derive a significant part of their identity from their engagement.

As a rule, members of sport subcultures tend to be more focused on activity and their sport community than on place (Hinch & Kono, 2017). Green and Chalip (1998: 275) captured this in their conclusion of their study

Table 6.1 Characteristics and types of participation in a distance running social world

	Outsiders	Occasional	Regulars	Insiders
Orientation	Naiveté	Curiosity	Habituation	Identity
Experiences	Disorientation	Orientation	Integration	Creation
Relationships	Superficiality	Transiency	Familiarity	Intimacy
Commitment	Detachment	Entertainment	Attachment	Recruitment

Source: Shipway (2008).

on women football players by stating that these women '…seek opportunities to share and affirm their identities as football players. It is the occasion to celebrate a subculture shared with others from distant places, *rather than the site itself*, that attracts them'. Participants were able to distance themselves from their regular lives, they enjoyed a sense of camaraderie with their teammates and other members of the subculture and they were given the opportunity to parade their subculture identities. The destination was described as facilitating the primary purpose of their sporting visit.

At another level, sport subcultures such as ultramarathoners are intimately connected and dependent on specific places for their sport (Hinch & Kono, 2017). Sites that facilitate the celebration of their sport identities and social worlds are highly valued. This is true of climbers, surfers, windsurfers, snowboarders and many other 'extreme' sport subcultures that are currently enjoying popularity. Traditionally, these groups tend to be very dependent on natural resources found in the periphery (see also Chapter 5). Their strong subculture commitment provides them with the motivation to overcome the constraint of distance. More recently, however, urban landscapes have become popular with these groups (Breivik, 2010; Wheaton, 2007, 2013). Subcultures and social worlds also exist within the fandoms of many urban-based professional sports. These groups also form strong attachments to place as evidenced by the emotional upheaval that they often experience when the traditional home ground is altered or abandoned.

It is common for the spaces used by sport subcultures to be contested. For instance, skiers did not welcome snowboarders when snowboarding was initiated on the slopes of the major European ski resorts (Heino, 2000). There may even be tension between the various types of members within a subculture as in the case of lifelong resident surfers, foreign resident surfers and tourist surfers in Costa Rica (Usher & Gomez, 2016). While the space associated with these sports may be shared, subculture place identity is often quite distinct.

Subcultural groups also recognise special places through access to 'insider information' (Donnelly & Young, 1988). As members of subcultural groups progress through their subculture careers they become privy to information about 'special' sites in the context of their group. The very act of travelling to selected destinations may garner social capital for an individual within his or her subcultural group (Shipway & Jones, 2007). Experience at these sites may be closely tied to a member's status within his or her subculture. Place can become one with the activity. Hardcore subculture members will tend to live near to where they can be active in their sport and will use their vacation time to travel to destinations that are 'sacred' to their sport. They are quite likely to develop travel careers that parallel their sport subculture career (Getz & McConnell, 2014).

Branding Place Through Sport

The tourism industry is in the business of selling places and this is done through the processes of marketing and place branding. While marketing has long been a focus of tourism practice and scholarship, the concept of 'place branding' has emerged as a popular term (Hanna & Rowley, 2008) that is closely associated with place marketing and place imaging. Marketers who try to shape the destination brand are usually trying to replace vague or negative images previously held by residents, investors and potential visitors (Page & Hall, 2003). In doing so, they are actively trying to influence place identity. Carter *et al.* (2007) suggest that place identity is often imposed by the globalising forces of development such as tourism. A good example of this is Urry's (1990) description of the 'tourist gaze' in which tourists judge their destination experience based on their image of the destination as drawn from marketing messages and more organic sources such as the popular media. There has been much debate about the tourism industry's ability to influence the tourist gaze (e.g. Garrod, 2009; Stepchenkova & Zhan, 2013), but sport's widespread popularity, the sporting activities found in a destination and a destination's reputation as a sporting site all represent promising resources for marketers trying to shape destination brands. The ability of marketers to influence destination images raises questions about the ethics of such manipulation and suggests that the promotion of place identity should be consistent with the sense of place prevalent in the community (Campelo *et al.*, 2013; Hinch & Holt, 2017).

Page and Hall (2003: 309) highlight the need to commodify particular aspects of place in the process of place branding. 'In the case of urban [or regional] re-imaging, marketing practices, such as branding, rely upon the commodification of particular aspects of place, exploiting, reinventing or creating place images in order to sell the place as a destination product for tourists or investment'. Sport is one of the most powerful ways of establishing a place brand given its cultural power. By harnessing these dimensions of sport, place marketers are able to commodify 'the ways of living' in a place. In a sport context, this can be accomplished with strategies related to: (1)

sport facilities, (2) sport events and (3) broad-based leisure and cultural opportunities within a destination (Hall, 1998).

Establishing a critical mass of visitor attractions and facilities has proven to be one of the most popular strategies for rebranding a city. Another high profile strategy for reimaging places is the hosting of events, particularly hallmark events (Getz & Page, 2016). Mega-events such as the Summer and Winter Olympics and the FIFA World Cup require huge investments which are often rationalised by their potential to position or brand the host city as a desirable destination and place to do business in a global world. This was the apparent goal behind Brazil's decision to host the 2014 FIFA World Cup and the 2016 Rio Olympics. In reality, there are significant risks associated with hosting these events. While such events may have positive impacts on the host city brand, they are likely to have negative impacts as well (Bodet & Lacassagne, 2012). Factors such as geopolitical events beyond the control of the local hosts may cause problems. Recent examples include the 2014 Sochi Olympics during which Russia had to deal with unrest in Chechnya (Reynolds, 2014). At the time of writing this edition, the 2018 Pyeongchang Olympics in South Korea were scheduled to open under the shadow of nuclear threats and military tensions between North Korea and the United States (Strashin, 2017). Given the high profile of such events and the critical nature of traditional and social media, any such issues are likely to receive considerable attention and may in fact leave lasting negative impressions on a global audience.

While there is a strong relationship between an event brand and a destination brand, they are not the same thing. Chalip and Costa (2005) identified three distinct roles that events can play in terms of the destination brand. Events can be: (1) co-branded with the destination brand; (2) positioned as extensions to the destination brand; or (3) situated as one of several features within the destination brand. Each option has its own advantages and disadvantages, but they need to be recognised and pursued based on a strategic assessment of the probable outcomes. Notwithstanding the challenges of using sporting events as part of the strategy for destination branding, the social dimension of these events and their potential to build social capital in the host community represent a major asset in building a positive destination brand. The destination brand will benefit if the host community is able to foster a genuine celebration in association with the event that leads to a state of communitas for event participants and the host community (Chalip, 2006).

Finally, the widespread development of sport-related leisure and cultural services is a third type of approach used to sell places based on sport. This approach goes beyond the support of high profile professional sports to the development of a sporting ethic within a destination through such initiatives as park and shorefront development that encourages active sporting pursuits. Lubowiecki-Vikuk and Basińska-Zych (2011) argue that all forms of physical activity including independent or casual activities have a bearing on the destination brand. These activities have impacts on a community's physical and mental health as well as social relations, all of which contribute to the ways that a destination is understood. Poznan, Poland, is an example of a city that has incorporated a sporting lifestyle into its destination brand in a number of ways including the reflection of its leisure-based character in its logo (Lubowiecki-Vikuk & Basińska-Zych, 2011).

Conclusion

This chapter highlights the importance of place in the context of sport tourism. Clearly, sport can have a powerful impact on the way that space is infused with meanings. Culture plays an important role in shaping these meanings whether it takes the form of sport and culture, sport as culture or sporting subcultures. Each variation impacts on the way that sport tourists see and experience a destination. Given sport's powerful influence on place identity, it is not surprising that it is often consciously manipulated to brand tourism places. Place marketers attempt to use spectator events, participation events, active sport, and heritage attractions to create a desirable place brand. One of the long-term challenges is for sport tourism destination managers to develop a sport brand that is consistent with the place identity of the locals. Failure to do so will test the sustainability of sport tourism in the destination.

A number of issues need to be considered in association with the many opportunities to brand 'place' through sport. Bale's (1989) spectre of the trend towards homogeneous sportscapes poses a threat to the critical elements of uniqueness, which in turn is a threat to the sustainability of sport tourism development. Taken to its extreme, in a homogeneous sportscape, the need or desire to travel to different areas for sport is greatly reduced. This is one of the challenges that is reflected in the process of glocalisation. While sport destinations strive to be part of a global sportscape, they need to preserve and even strengthen their local sport culture to achieve success in this area. These tensions between the local and the global are particularly well illustrated in the case of Liverpool FC (Case Study 6.1). Fundamentally, the integrity of the sports being commodified for sport tourism must be protected. Place marketers must avoid the temptation to sensationalise or spectacularise featured sports in a way that erodes the essence of the sport competition and the local destination. While place marketers who seek to develop powerful destination brands are in the business of commodifying both place and sport, if the local meanings associated with sport are compromised or destroyed, then the sport resource and the attractiveness of sport as a tourist attraction will also be compromised.

Finally, it must be recognised that there are multiple meanings of place in tourism spaces (Schollmann *et al.*, 2001; Sherlock, 2001). Place marketers who use sport as a marketing tool need to appreciate the contrasting perspectives of place held by different groups within the community. Distinct place meanings are not just associated with hosts and guests but also with the complex array of subgroups that exist therein (e.g. in the case of alpine resorts: long-term and short-term residents, second home owners, skiers, snowboarders, climbers and numerous others). Failure to account for these differences may result in conflicting views on place, which are sub-optimal at the very least, and may in fact be non-sustainable in the long run.

7 Environment: Landscape, Resources and Impacts

Sport tourism's link to the environment is both as victim and as aggressor.
Standeven and De Knop, 1999: 236

Introduction

Sport-related tourism development is tied more closely to the geographical resource base of a destination than many forms of tourism. The extent to which tourists find a destination attractive is strongly influenced by the physical environment including landscapes and climate (Boniface & Cooper, 1994; Hall & Page, 2014; Krippendorf, 1986). Many sports are closely tied to the physical geography of a destination. For instance, Priestley (1995: 210) observes that single integrated golf resorts '...have mushroomed in the hotter climates where traditional sun, sand and sea tourism could or does exist'. In sports such as surfing, hang-gliding and scuba diving, there tends to be a hierarchy of destinations based on the experiential value of the physical environment. Destinations may be managed and promoted to develop new or exploit existing links to specific sports. The development of integrated golf resorts in Spain capitalises on increasing levels of visitor demand for this sport (Priestley, 1995), in conjunction with the hypermobility of European nationals, particularly with the growth in demand for low-cost air travel as an increasingly instituted social practice (Randles & Mander, 2009). Similarly, the confluence of serious leisure and competitive mountain biking has been recognised and harnessed in different ways in the development strategies of a number of tourism destinations in New Zealand (Moularde & Weaver, 2016).

The sport tourism development potential of a destination is also determined by cultural influences on the landscape. Event sport tourism development at a destination requires, in most cases, constructed resources including sport facilities and tourism infrastructure. Sports in central locations often use facilities that are purpose built, such as stadiums, marinas, sports arenas and gymnasiums. Alternatively, sports may temporarily make use of buildings or infrastructures that are developed primarily for purposes other than sport. Examples include roads, central parks and urban tourism icons, which may figure prominently as locations or backdrops (e.g. the statue of Christ the Redeemer during the Rio 2016 Olympic Games) to sporting scenes (e.g. Copacabana Beach; venue of the 2016 Olympic beach volleyball competitions). Gilchrist and Wheaton (2011) provide intriguing insights into how the non-traditional, non-institutionalised lifestyle sport of parkour has, without any dedicated facility development whatsoever, engaged communities of participants, many of whom are otherwise excluded from involvement in mainstream sports. An understanding of the spatial elements of sport tourism development is therefore incomplete without some consideration of the physical environment. This is an important starting point to understanding the resource requirements and impacts of sport tourism development. Natural and built resources for sport tourism, and impacts associated with each, are considered separately in this chapter.

Sport Tourism Landscapes, Environments and Resources

'Landscape' is a term that is commonly associated with attractive scenery. Natural landscapes (and seascapes) are central to the pursuit of many sports. However, sports are not natural forms of movement and, therefore, 'the landscape upon which such body culture takes place is part of the cultural landscape' (Bale, 1994: 9). Even sports that rely on natural elements take place in environments that are subject to varying degrees of anthropogenic change. For instance, ski slopes are subject to change through the grooming of ski trails, the construction of facilities such as ski jumps and slalom courses, snowboard half pipes and ramps, snow-making and the development of visitor services (Hudson & Hudson, 2010). Golf courses, which are very 'green' in appearance, represent highly modified natural areas and are characterised by significant ecological impacts (Briassoulis, 2007; Rodriguez-Diaz et al., 2007; Wheeler & Nauright, 2006) (Focus Point 7.1).

Focus Point 7.1: Water Resources and Golf

Michelle Rutty

Golf tourism has grown rapidly in the last two decades, drawing considerable negative attention to the sport. Key environmental concerns include long-term impacts associated with the use of pesticides and chemicals on human and wildlife health, large-scale modification of the natural environment and associated ecosystem impacts and the high consumption of freshwater resources for course irrigation (e.g. Briassoulis, 2010, Wheeler & Nauright, 2006). The latter is arguably the most contentious, with several media campaigns contesting the water-intensive nature of the sport, resulting in water resource conflicts in several regions (e.g. Harvey et al., 2013, Priestley, 2006). Despite the negative attention, water footprint analyses for golf tourism remain limited.

(Continued)

Focus Point 7.1: (Continued)

In 2004, it was estimated that 9.5 billion litres of water were used daily to irrigate the world's courses (Wolbier, 2004). Although the robustness of this estimation has been questioned, the magnitude of the issue is apparent. Both regionally and intra-regionally, water use for golf course irrigation has been found to vary substantially (for an overview, see Scott *et al.*, 2018). For example, a typical 18-hole golf course in the US Northeast uses an average of 52 million litres annually versus 566 million litres in the Southwest (Throssell *et al.*, 2009). In the Mediterranean, Portugal's courses use 438 million litres (Videira *et al.*, 2006) versus up to 1 billion litres in Cyprus (Meulen & Salman, 1996). In the UK, water use varies from 15 to 97 million litres (Environment Agency, 2008) and in Dubai, between 1.36 and 2.68 billion litres (Todorova, 2015). Water consumption is heavily dependent on several factors, including course size, design, management, soil characteristics and climate. The lack of current water use profiles for courses precludes an ability to compare golf tourism's annual water use (and economic value generated per unit of water) to other water-intensive industries, both within and outside of tourism.

Although a complete account of golf's water footprint still must emerge, measures that reduce water extraction rates are necessary next steps to secure golf's future in water-constrained regions, and in a world of increasing freshwater scarcity. The sport's largest governing bodies are responding, developing innovative water use practices that reduce the need to draw so heavily on freshwater resources (e.g. Golf Canada, 2014; Royal and Ancient Golf Club of St. Andrews, 2010; United States Golf Association, 2014). Best management practices include the incorporation of drought-tolerant turf grasses and/or native vegetation that are acclimatised to local climatic conditions, applying wetting agents to increase soil moisture for extended periods of time, harvesting rain and storm water to reduce extraction rates from freshwater resources, using soil moisture sensors to reduce overwatering, hand-watering target areas most in need and restricting irrigation to the fairway (i.e. not watering the rough and out-of-play areas). The golf industry also has the potential to influence perceptions of golf course ideals and engage golfers in water-savings opportunities, such as shaping expectations of turf aesthetics, which is an often-overlooked water savings opportunity.

Key Reading

Scott, D., Rutty, M. and Piester, C. (2018) Climate variability and water use in golf tourism: Optimization opportunities for a warmer future. *Journal of Sustainable Tourism (in press)*.

Additional references cited in this focus point are included in the reference list.

While the popular use of the term 'landscape' often implies naturalness, the landscapes of sport are, to varying degrees, cultural landscapes. The term 'sportscape' is used in the geography of sport to describe the highly modified (e.g. modern stadium or arena) and technologised (e.g. video replay screens, video referee systems) sports environment (Bale, 1994). Relph (1985: 23) notes that landscapes can 'take on the very character of human existence. They can be full of life, deathly dull, exhilarating, sad, joyful or pleasant'. This observation certainly applies to the landscapes of sport. The manner in which the landscapes of sport are developed, the levels or forms of engagement and the impacts arising from the use of those landscapes are important to the sustainable development of sport and tourism.

The Landscapes of Sport

The values and interpretations associated with the landscape are highly subjective (Tuan, 1977). Sportscapes are no exception. Bale (1994) applies Meinig's (1979) 'ten versions of the same scene' to the landscape of sport in an exercise that is relevant to the study of sport tourism (Table 7.1). These 'versions' are important in understanding the resources and impacts of sport tourism. The development of resources and infrastructures for sport tourism should take place with consideration given to the values and interpretations of landscapes that are outlined and explained in Table 7.1.

The resource base for sport tourism

The potential for sport tourism development at a destination is determined by the existence of requisite sport and tourism resources and infrastructures. A sport tourism resource analysis may include natural environments, constructed sports facilities, tourism transport and infrastructure and information services. These need to be provided in the required balance and combination, or developed in a planned and coordinated way as determined by the development goals of the destination (Maier & Weber, 1993). The importance of coordinated planning and development arises from the considerable overlap that exists between the resource requirements for sport and those for tourism (Standeven & De Knop, 1999). Domestic and international airline services are used by travelling sports teams and leisure travellers for the same purpose, whereas both use stadiums, albeit for different reasons (competition and spectatorship, respectively). The existence or systematic development of sport and tourism infrastructures is required for any location to function as a sport tourism destination (Table 7.2).

Considerable opportunity exists, therefore, for sport and tourism resources to be developed in a synergistic fashion that maximises mutual sport and tourism and wider social and sustainable development interests.

Table 7.1 Interpretations of sport landscapes

(1) Sport, landscape and natural habitat	It is possible for sport participants to encounter and utilise the natural landscape for certain sports events and, when the event is over, never return to it. They remain landscapes and never become sportscapes. Landscapes therefore may be used for sports but never 'sportised' in any permanent sense. Impressions of nature and environment are important elements of the athlete's experience.
(2) Sport, landscape and human habitat	The sport landscape may also be regarded as part of the human habitat. Conscious decisions can be made for slopes, soils, elevations, sites and routes, channels or relief features to be used as homes for sport. Humans rearrange nature into sport-related forms; an adjustment rather than a conquest of nature.
(3) Sport landscapes as artefacts	Many sport landscapes disregard the natural or semi-natural landscape upon which they are found. This view sees humankind as the conqueror of nature, with concrete, plastic and glass, totally flat synthetic surfaces and indoor arenas in which nature has been neutralised.
(4) Sport landscapes as systems	Sports landscapes can also be viewed as part of intricate economic or physical systems. A sports stadium, for example, does not exist in isolation; it generates flows of people and spatial interactions over an area much greater than that of the stadium itself. For example, the Tour de France is part of an extensive economic system that affects the places through which it passes. Sports events are also part of physical systems. Snow conditions influence performance in ski races and rain may deter attendance at sports events.
(5) The sport landscape as problem	The excessive dominance of sport over nature may be seen to lead to social or environmental pollution, erosion and visual blight. Problem landscapes occur in a variety of sports in quite different ways. Traffic congestion and crowding can result from hosting a sports event in an inner-city stadium. Erosion of soil and damage to plant cover on ski pistes in alpine regions are also examples. Impacts also differ in terms of their permanence. When the sports landscape becomes perceived as a problem, it can lead to political activism and the rejection of sporting events that might have induced landscape change.
(6) Sport landscape as wealth	The sports landscape may also reflect the view that land is a raw material. The long-term returns of lands given over to sport are important. So, too, are the significant economic benefits that one-off events generate in local areas. Sport may be a form of place-boosting for purposes of attracting investment, and may influence rental profits. The sports landscape is littered with advertising hoardings and other evidence of sponsorship.
(7) Sport landscape as ideology	The sports landscape may be viewed as a reflection of various ideologies. Sports landscapes may be explicit responses to nationalism. New national sports may be invented to distance countries from their more dominant neighbours. The stadium may be an expression of modern technocentric ideology.
(8) Sport landscapes as history	The present-day landscape of sport is a result of the cumulative processes of historical evolution. Sports landscapes are often accumulations. Size, shape, materials, decorations and other manifestations tell us something about the way people have experienced sport over time.
(9) Sport landscape as place	This view sees landscape as a locality possessing particular nuances and unique flavours that express a unique sense of place. For the sports participant – athlete or spectator – the experience of place, therefore, could be argued to contribute to the overall sporting experience.
(10) Sport landscape as aesthetic	Landscapes can possess aesthetic qualities predisposing the observer towards one and against another. Aesthetics are related to the artistic quality of the sports landscape. The aesthetics of the sport landscape are also portrayed in paint, film, photograph and print. Such portrayals may be accurate representations of what exists in the physical landscape. It is also possible that landscape icons may become mythical landscapes (e.g. the landscapes of English cricket and American baseball).

Source: Adapted from Bale (1994).

Table 7.2 Resource-base for sport tourism development

Tourism industry resource requirements	Sport sector resource requirements
Natural features	Natural features
National parks, scenery, lakes, mountains, rivers, coastlines	National parks, open amenity spaces, wilderness areas, geographical features (mountains, rocks, spas, coastlines, marine environments)
Facilities and infrastructure	Facilities and infrastructure
Transport services, places of accommodation, dining and entertainment	Stadiums, arenas, sports halls, transport infrastructure, dining and entertainment
Built amenities	Built amenities
Public toilets, parking facilities, signposts, shelters	Public toilets, parking facilities, signposts, shelters
Tourist information services	Sport services
Visitor information services, internet-based information services, booking and ticketing services, travel agents	Coaching and leadership, equipment/clothing hire and/or purchase, storage and management, supervision and safety, hiring, operations, training facilities, injury prevention and medical facilities, science and research facilities
Tourism organisations	Sport organisations
Planning and development, strategic planning, destination image, tourism marketing, place promotion, visiting media programmes, tourism research, industry coordination and liaison	Sports clubs, volunteer groups and community groups; administration, facility development, funding, sponsorship, information services, marketing, merchandising
Transport services	Transport services
Road, rail, air, sea, domestic and international. Plus scenic journeys, gondolas, tourist routes, rides, heritage rail tourism, historic routes, tour coaches, hot air balloons.	Road, rail, air, sea, domestic and international
Entertainment and activities	Entertainment and activities
Attractions, casinos, cinemas, zoos, shopping, nightlife, nightclubs	Sports halls and venues (ice rinks, leisure centres, gymnasiums, swimming pools, climbing walls), golf courses, marinas, sports museums, halls of fame, shopping, nightlife

Source: Adapted from Standeven and De Knop (1999).

Event sport tourism, for example, offers the potential for the inner-city resource base for sport, recreation, entertainment, retail and service to be transformed in a planned and coordinated manner. This course of strategic development may generate the advantages of enhanced profile and destination image vis-à-vis sport tourism, thereby improving the standing of the destination in the hierarchy of sport tourism locations (see Chapter 5). The status of a ski destination, for example, is a function of high-quality ski resources (e.g. terrain, elevation, snow conditions, weather), in combination with the required tourism services and infrastructure.

The reproducibility of sports

The tourism resource base may be classified in various ways. One approach draws on the distinction between those that can be reproduced, or transported, and those that are non-reproducible (Boniface & Cooper, 1994). Resorts, theme parks and stadium experiences are readily reproduced and can be developed in, or transmitted to, a variety of locations (Weed, 2010). In contrast, natural landscapes and cultural heritage are generally non-reproducible, despite efforts to the contrary. Sports resources may also vary on the basis of their transportability. Nature-based sports such as downhill skiing and rock climbing tend to be dependent on certain types of landscapes or specific landscape features. Attempts to create artificial ski slopes in central locations have met with moderate commercial success (see Chapter 10). The reason for this is that the experiential value of the mountain environment, which forms an important part of the participant experience for many sport tourists, is not transportable (although spectators may dramatically experience alpine skiing through film media, such as the Banff Film Festival).

The same is true of indoor climbing walls (Kulczycki & Hinch, 2014). While they present an exciting variation of rock climbing, they cannot duplicate the unique challenge of outdoor climbing sites. Nature sports are those that are dependent on the integration of a physical activity with specific environmental attributes (Bale, 1989). Sports such as surfing, cross-country skiing, windsurfing, sailing, mountain climbing and orienteering are examples of green sports, as they are built around specific features of the natural environment as sources of pleasure, challenge, competition or mastery. A case in point is the way that hang gliding, parapenting and windsurfing harness the natural forces of air and sea. As a result, participants enjoy a heightened sense of environmental awareness due to the role it plays in their performance. Embodiment is a critical element of nature-based sports (Humberstone, 2011; Lamont, 2014). The experiential value of these sports is largely dependent upon the mood of the landscapes where they are performed; a combination of landscapes and terrain (including scenery), climate and weather. These landscapes are inherently non-transportable. In contrast, other sports are more readily transported. Indoor sports such as ice-skating have been successfully transported from high to low latitudes with the development of improved ice-making technology and expanding markets. Indeed, indoor arenas have transformed sports such as ice hockey from outdoor to indoor activities impacting their spatial and temporal distribution (Higham & Hinch, 2002a), allowing sport to survive in a changing climate, and giving some certainty to the viability of major sports events (see Steiger & Abegg, 2018; Case Study 9.1). Spatially, these sports have dispersed from their places of origin in the high latitudes and temporally from highly seasonal winter sports to year-round activities. Outdoor winter sports such as ski jumping may also be transported from peripheral to central locations in the high latitudes to capture the advantage of proximity to markets. The Holmenkollen (Oslo, Norway) and Vancouver 2010 Olympic (Canada) ski jumps are examples of constructed ski jump facilities that have been developed in or close to central locations.

Many sports, such as competitive swimming, diving, squash and racket ball, are performed in indoor sports centres and have become very transportable. These sports are also characterised by highly prescribed spatial rules and standards. Other sports that are traditionally played in outdoor settings can also be transported and performed in indoor sports centres and arenas. Examples now include tennis, athletics, football, rugby and equestrian activities. These sports demonstrate what Bale (1989: 171) refers to as the 'industrialisation of the sport environment', which relates closely to the concept of transportability. Indoor cricket, for example, is a sport that takes place in air-conditioned centres that are typically housed in unused industrial buildings and warehouses located in industrial landscapes (Bale, 1989).

The application of technology to the modern stadium demonstrates the height of sport transportability. The reproducibility of the sportscape facilitates the transportation of sports and the sport experience. Rugby is an example of a transportable sport, which offers opportunities for generating new markets and revenue by hosting international test matches in neutral countries/cities. In recent years, the New Zealand All Blacks have played test matches in Tokyo, Hong Kong (versus Australia) and Chicago (versus Ireland) to break into new and emerging spectator and media markets. Viewed another way, sports facilities may be built, permanently or temporarily, at locations designed to maximise market access. Such developments offer the potential to enhance the status of sports, such as snowboarding and beach volleyball, through increased public awareness and spectatorship. However, the transportability of sports also presents the threat of the displacement of a sporting activity from its original location. The importance of retaining and enhancing the idiosyncrasies and elements of uniqueness associated with a tourism site is an important strategy to mitigate this threat (Bale, 1989).

Environmental Impacts of Sport Tourism

From a geographical perspective, 'the environment is the totality of tourism activity, incorporating natural elements and society's modification of the landscape and resources' (Mitchell & Murphy, 1991: 59). An understanding of the impacts of sport tourism, and management techniques appropriate to those impacts, is central to the sustainable development of sport and tourism. 'Inevitably, the growth and continuing locational adjustments made by modern sports have created significant changes in the landscape' (Bale, 1989: 142). Many such impacts are fleeting, or temporary. Triathlons, marathons, cycle races, car rallies and festival or exhibition sports are often conducted on circuits, courses or courts that may be constructed temporarily in urban areas. The impacts of these sports, which may include a sizeable body of spectators, are rapidly dispersed at the conclusion of the event. The immediate negative consequences of stadium-based sports may include traffic congestion and crowding, and undesirable impacts such as vandalism, antisocial behaviour, littering and noise. These impacts are generally short term but they can cause disruption to community residents (Bale, 1994). They may also result in aversion effects upon visitor flows into or within a destination as non-sport tourists choose to visit other destinations or cancel intended visits during scheduled sport events (Weed, 2007).

Other sports may have a longer-term or indelible impact in cases where naturalness forms an important, perhaps central element of the sport tourist experience. While environmental impacts in natural areas may be 'permanent but, paradoxically annoying to few' (Bale, 1994: 11), one significant consequence may be a compromise of the quality of the sport tourist experience. This is illustrated by the stagnation of ski markets in Europe and North America as a demand-led response to the unsustainable management of environmentally sensitive alpine environments (Flagestad & Hope, 2001). Indeed, the complex interrelationships that exist between sport and tourism are particularly evident in some of the global forces of environmental change that are now influencing the quality and indeed viability of the sport and tourism natural resource base (Marshall *et al.*, 2011; Pickering *et al.*, 2010; Scott & McBoyle, 2007).

Sport tourism and globally dispersed environmental impacts

The environmental impacts of sport tourism now require that consideration extends beyond impacts that are local/regional in scale and concentrated in nature, to the forces of anthropogenic environmental change that are global and dispersed. Sport tourism, by its very nature, involves the generation of significant one-off and recurring flows of tourists (Weed & Bull, 2012). These flows include elite athletes, who engage in constant travel in search of competition, as competitive leagues have become increasingly globalised (Higham & Hinch, 2009). They also extend to new and intense mobilities of leisure, recreation, sport and tourism. Recognising and responding to the drivers of global environmental change is a critical challenge (Hinch *et al.*, 2016). One of the most pressing challenges relates to sustainable tourism transportation and the need for urgent low carbon mobility transitions (Hopkins & Higham, 2016).

While transportation accounts for 23% of total global energy-related CO_2 emissions, transport sector emissions are projected to double by

2050 (Creutzig *et al.*, 2015). Aviation and vehicle use produce a large and growing portion of the world's greenhouse gas emissions (Bows-Larkin *et al.*, 2016; Creutzig *et al.*, 2015). The urgent need to address transportation emissions was recognised in the Paris Climate Accord (*L'Accord de Paris*), which carries the commitment of 196 countries (Parties to the Agreement) to the overarching goal of stabilising global average temperatures below +2°C relative to pre-industrial levels (UNFCCC, 2015), with many signatories committing to pursue a 1.5°C target (Scott *et al.*, 2016a, 2016b). It is now incumbent upon signatory nations to develop policies to meet the national determined contributions (NDCs) that they committed to in Paris in December 2015. The implications for the sport and tourism sectors will be inescapable.

The sport and tourism sectors must pursue emissions mitigation, given the critical and complex interrelations and interdependencies that tie sport and tourism to the environment. A key challenge in sustainable sport and tourism development has emerged not only in relation to the movement of tourists from generating to destination regions to engage in sports, but also in terms of the greenhouse gas emissions associated with movement of sport participants and spectators (Heath & Kruger, 2015). Scholars have recognised the ironies that exist in relation to nature-based sports and high participant mobilities. Hopkins (2014), for example, observes the irony that tourists who travel internationally and domestically to engage in alpine skiing in New Zealand contribute significantly to the transport emissions that are driving climate change which will ultimately destroy the winter ski resource (Case Study 7.1). Hinch *et al.* (2016: 165) consider that '(transport) emissions may ultimately become part of the energy inventory and emissions profile of the destinations that host sports events and benefit from the flows of tourists'. There is no doubt that questions relating to the sustainability of sport tourism must be reframed to account for the drivers of global environmental change, in which sport and tourism are implicated.

Case Study 7.1: Winter Sports Resources, Climate Change and the Ironies of Sports-Related Mobilities

Debbie Hopkins, University of Oxford, UK

The popularity of winter sports tourism has ebbed and flowed over the past century, through periods of mass growth in popularity and investment in infrastructures, to periods of consolidation (Hudson, 2003). Events such as the Winter Olympics and the X Games, and the emergence and growth of new winter sports, have helped to increase demand among adolescents and young adults, and broadened the image

(Continued)

Case Study 7.1: (Continued)

of winter sports from functional and traditional activities such as snow shoeing, cross-country skiing and ice skating, to board-cross, half pipes and backcountry activities. Yet, globally, the snow sports industry has entered another challenging period – and changing weather patterns associated with climate change are at its heart.

Snow sports require specific and stable weather and climate conditions (Hamilton *et al.*, 2007). Small changes in humidity, temperature and wind patterns can have lasting effects on the profitability of winter sport tourism destinations – particularly ski fields and their infrastructure (Hendrikx & Hreinsson, 2012). The vulnerability of the ski industry as a subsection of the snow sports industry relates to its place dependence, high infrastructure requirements and limited organisational adaptability. Adaptation to climate variability has been largely centred on snow-making technologies (Hopkins, 2014), which have been used to replace natural snow, 'top up' natural snow and/or extend the winter season. However, in some regions, diversification into non-snow-reliant activities such as downhill mountain biking has been an important adaptation strategy, particularly where the local climate is not conducive to snow-making (Hopkins & Maclean, 2014).

The impacts of climate change are threatening the sustainability of winter sports tourism across the globe, from traditional ski regions including Canada (e.g. Rutty *et al.*, 2017) and Europe (e.g. Steiger & Stötter, 2013), to non-traditional destinations such as Iran (Ghaderi *et al.*, 2014). While the impacts are not uniformly experienced, few destinations are not being affected to some degree, either as 'winners' or 'losers'. Research has shown that the relative vulnerability of ski fields, regions and countries could be more important than individual vulnerability assessments. In other words, vulnerability needs to be understood relationally (Hopkins *et al.*, 2013). Snow sports *tourists* have greater adaptability than the tourism operators, and thus have the ability to change geographical location locally or globally, or to change activity – for instance by opting for a 'winter sun' holiday. The behavioural adaptation of snow sports tourists (e.g. Cocolas *et al.*, 2016; Dawson *et al.*, 2011) contributes to the declining profitability of ski fields, with implications for the wider, often rural, communities. And while snow sports were once a domestic leisure activity, the global extension of snow sports tourism flows, and related dependence on air travel, means that the mobilities of winter sport tourists will contribute substantially to climate change. Thus, the (mal)adaptability of snow tourists is likely to result in greater mobility to access snow, more long-haul air travel and more emissions, thereby increasing the impacts of climate change.

There are significant and problematic ironies of snow sport-related mobility. Both tourism in general, and transport as a particularly high-emitting activity, contribute to global greenhouse gas emissions, the main cause of anthropogenic climate change (Scott *et al.*, 2016a). Ski fields are resource intensive – they require both energy and water resources to maintain operationality. Some ski fields have developed renewable electricity resources, harnessing wind- or hydroelectricity to power lifts, buildings and other facilities. Yet, contestations between renewable electricity infrastructure and ski lifts and cable cars have been reported (Huber *et al.*, 2017). Furthermore, with the growth of snow-making technologies, access to water resources has become of increasing importance. This has led to demand for high-elevation reservoirs and rights to access water resources. Transport-related emissions, both internationally and domestically, are also highly problematic in light of climate change and its impact on the ski industry. With often rural, alpine locations, ski fields are frequently accessed by private vehicle, and although some ski fields have been successful in implementing public transport systems and ride-sharing schemes, ground transport remains a neglected area of ski tourism sustainability management (Rutty *et al.*, 2014).

Transport-related greenhouse gas emissions have more than doubled since 1970, and the growth in demand is outweighing efficiency gains (Sims *et al.*, 2014). The Paris Climate Agreement set forth an ambition to stabilise the global climate at 1.5°C above pre-industrial times, yet once again, global aviation was omitted from inclusion (Scott *et al.*, 2016b). The aviation industry has failed to take meaningful action on its greenhouse gas emissions. Similarly, the tourism industry has been relatively weak in its response to climate change (Hopkins & Higham, 2018). Nevertheless, there have been some efforts to raise the profile of climate change. For instance, the 'Keep Winters Cool' campaign, a joint effort by the Natural Resources Defense Council and the National Ski Areas Association (Scott, 2006a, 2006b), emerged in the United States but has since gained traction globally, thanks, in part, to the involvement of high-profile snow sports athletes and a strong social media platform. The engagement of professional athletes with the issue of climate change relates not only to their experiences of a changing climate, but also to financial constraints. For example, 'a few bad winters' and 'shifting weather patterns' have been blamed, along with the weakened global economy, on the scarcity of sponsorship deals for professional snow sports athletes, and a 'challenging atmosphere' for snowboarding in particular (Higgins, 2016).

(Continued)

Case Study 7.1: (Continued)

The winter sports industry is heavily reliant on stable climatic features and natural resources. Winter sports are highly place dependent, and often have little by way of adaptability. Over the coming decades, many popular ski destinations are likely to become unprofitable and the environmental sustainability of these resorts is declining. Traditional centres of snow sports tourism may shift as the impacts of climate change increase. However, these are likely to be even more dependent on long-haul air travel – further exacerbating global climate change. In order to address the current complex problems of high mobility, high emissions and accelerating climate change, there is a critical and urgent need for a low carbon mobility transition (Hopkins & Higham, 2016). The success of the Paris Climate Agreement and systemic changes to high polluting industries, including tourism, will be critical to protecting the climate system upon which winter sports destinations are so dependent.

Key Reading

Cocolas, N., Walters, G. and Ruhanen, L. (2016) Behavioural adaptation to climate change among winter alpine tourists: An analysis of tourist motivations and leisure substitutability. *Journal of Sustainable Tourism* 24 (6), 846–865.

Hopkins, D., Higham, J. and Becken, S. (2013) Climate change in a regional context: Relative vulnerability in the Australasian skier market. *Regional Environmental Change* 13 (2), 449–458.

Additional references cited in this case study are included in the reference list.

Sport Tourism in the Built Environment

Much of the existing literature on the environmental impacts of sport tourism has focused on natural areas (Standeven & De Knop, 1999). However, sport tourism development in urban areas presents unique environment, resource and impact issues, which require informed consideration. Sport tourism in the urban context may include:

(1) Active sport and physical exercise in the built sports landscape (leisure centres; hotel gymnasiums; squash, badminton and tennis courts; swimming pools), recreational running in urban parks and developed littoral zones.
(2) Recreational or club sport in dedicated sports fields or improvised settings (e.g. skateboarding and street basketball).

(3) Recreational or competitive sports that take place in a largely unmodified (e.g. kayaking, surfing) or reproduced nature (e.g. orienteering in an urban conservation reserve).
(4) Event sport tourism.

Active, recreational and competitive sports in the urban context generate relatively benign impacts although they may require management of social impacts or conflict between participants and non-participants. Sport events, in the form of elite or non-elite sports that require dedicated or temporary facilities, which take place in central locations offer considerable potential for both positive and negative impacts. These impacts are a function of the scale of the event and the infrastructure capacities of the destination.

Issues of scale in event sport tourism

Scale, be it global, regional or local, is critical to the study of sport tourism in central locations. 'The idea of scale, or geographical magnitude, keeps in focus the area being dealt with, and can be likened to increasing or decreasing the magnification on a microscope or the scale of a map' (Boniface & Cooper, 1994: 3). The capacity for locations to accommodate flows of tourists is determined in large part by the scale of the destination, and its capacity to absorb tourists. The Tourist Function Index, for example, employs the number of tourist beds and the total resident population of a destination as indices of tourist capacity (Saveriades, 2000). The concept of tourism carrying capacity considers the maximum level of tourist activity that can be sustained without adversely impacting the physical environment or the quality of the visitor experience, with consideration given to the views of the host community (Archer & Cooper, 1994; Wall & Mathieson, 2006).

'[M]ega-events are short term events with long term consequences for the cities that stage them' (Roche, 1994: 1). Unfortunately, interests in the impacts of event sport tourism are often restricted to economic development (Burgan & Mules, 1992), positive image and identity, inward investment and tourism promotion (Getz, 1991; Hall, 1992a, 1992b). This focus ignores the potential for sporting events to create negative impacts, which tend to increase with the scale of the event (Olds, 1998; Shapcott, 1998). Where the scale of a sports event is too great for the social and infrastructure capacities of the host city, significant potential for negative impacts arise (Hiller, 1998). Host community displacements and evictions (Olds, 1998), increases in rates and rents (Hodges & Hall, 1996), the disruption of daily routines due to crowding and congestion (Bale, 1994), security issues (Higham, 1999) and the exaggerated behaviour of 'sports junkies' (Faulkner *et al.*, 1998) may be associated with large-scale sporting events. Shapcott (1998), for example, reports that

> 720,000 room-renters (were) forcibly removed in advance of the 1988 Olympics in Seoul, thousands of low income tenants and small businesses forced out of Barcelona before the 1992 Games (and) more than 9000 homeless people (many of them African-American) arrested in the lead-up to the 1996 Olympics in Atlanta. (Shapcott, 1998: 196)

Sporting events and competitions of more modest scale include regular season domestic sport competitions, national/regional championships and non-elite sports events. At these more modest scales, the potential for serious negative impact is reduced (Higham, 1999). Crowding and infrastructure congestion are less likely to occur and are more rapidly dispersed. Nonetheless, the positive impacts of sports events of more modest scale are, within the geographical parameters of the destination, very similar to mega-events (Gibson *et al.*, 2012; Hall, 1993), although easily oversold (Gratton *et al.*, 2005a; Whitson, 2004). Issues of scale in event sport tourism are critically important. Although the economic imperative that 'bigger is better' remains prevalent (Weed, 2009), the achievement of a match between the capacity constraints of host cities or regions and the scale of the sports events that they seek to host, represents an important element in achieving sustained and sustainable success in event sport tourism (Gratton *et al.*, 2005a; Higham, 1999).

Managing the compatibility of sports in the built environment

Consideration of the compatibility of multiple sport demands is an important sport tourism management issue in the built environment. Different sports may be viewed as:

(1) Compatible: Sports that can use the same area of land or water at the same time.
(2) Partially compatible: Sports that can use the same area of land or water but not at the same time.
(3) Incompatible: Sports that cannot use the same area of land or water and need to be zoned into exclusive spaces.

The extent to which different sports demonstrate compatibility with other landscape users varies considerably. For example, motor sports and sports involving dangerous equipment (e.g. field archery) are essentially incompatible with other sports. The incompatibility of sports generally increases at higher levels of competition. Competitive or elite levels of sport require specialised and sometimes exclusive use of facilities. The sports manager, then, must be mindful of the required balance between specialisation and multiple use in the design of sports facilities. Following the lessons learned in Montreal (1976), the Sydney (2000) and London (2012) Olympic stadiums were designed for post-event downsizing for subsequent use.

The development of multiple-use facilities, particularly those that cater for sports at various levels of competition (ranging from local/recreational to international/championship), may diversify and expand the user and spectator market catchments for a facility. Consideration should be given to both the spatial (e.g. dimensions of the playing surface, parking and spectator capacities) and temporal (e.g. daily/week use patterns, sport seasonality) compatibility of sports that may derive mutual benefit from the use of a multiple-use facility. In some cases, however, the development of generalised or multiple-use facilities can cause unacceptable compromises to the sport experiences of both participants and spectators. Stadiums with running tracks, for example, typically are characterised by non-optimal viewing for a high proportion of spectators (Bale, 1989).

The issue of compatibility in the built landscape extends to reconciliation of sport/non-use interests, particularly at sites that are designed primarily for purposes other than sport. For example, the marathon and distance running boom, with race fields exceeding 20,000 participants in some cases, has '...put pressure on municipal authorities to control and redirect traffic on race days' (Bale, 1989: 163). Streetcar rallies, cycle races and a host of festival sports may also cause disruption to the normal use of the urban landscape. Such impacts tend to be short term and rapidly dispersed. They do, however, require that sport managers and event organisers consider security, safety and liability issues relating to their sport.

Landscape to sportscape: The impacts of sport facility development and design

It has been noted that 'the search for regional diversity in the landscape has remained an important motive for travellers, despite the standardisation and homogenisation of the tourism industry' (Mitchell & Murphy, 1991: 61). There exists an evolutionary tendency to confine and homogenise the sporting environment. A transition from landscape to sportscape represents one aspect of standardisation and homogenisation in sport tourism, which may seriously threaten the uniqueness of specific places. The modern stadium has an ancient history, and has evolved through phases that have been influenced by the formalisation of sports rules and the imposition of spatial limits in sport, which allowed the development of facilities for spectators to experience sport contests at close proximity (Bale, 1989).

More recently, technological developments, such as video screens, virtual advertising, floodlighting and retractable enclosures, have been incorporated into modern stadium design (Bale, 1989). This course of development has given rise to an increasing sameness of stadium design in many parts of the world (Higham & Hinch, 2009), which may significantly alter the overall sporting experience from the viewpoint of both competitors and spectators. One implication may be the erosion of 'the cultural mosaic that encourages tourism' (Williams & Shaw, 1988: 7). The potential contribution to sport tourism development of unique stadium design, contiguous markers, distinctive elements of the destination and the natural elements that differentiate destinations must be carefully considered in relation to the planning, design and development of sports resources.

The counter trend, which rejects the standardisation of the sportscape, is evident in some cases, with important implications in relation to uniqueness, differentiation, sense of place and tourism. When the famous Wembley Stadium (London), along with its distinctive towers, was demolished in 2002–2003, a piece of sporting heritage was lost. The rebuild of Wembley Stadium included a retractable roof and the distinctive 134 m high Wembley Arch, which is visible across the London cityscape. When South Korea co-hosted the *Internationale de Football Association* (FIFA) World Cup in 2002, the medium of stadium design was used to project Korean culture and way of life. Each of the 10 host cities in Korea (re) developed stadiums that were designed to reflect the important elements of regional culture that are unique to each city (Hinch & Higham, 2004). Munhak Stadium (Incheon), for example, has a sailboat-designed roof to reflect the city's historical role as Korea's leading maritime gateway, while the traditional curved roofs of the Suwon Stadium symbolise the ancient Hwaseong stone fortress, which dates back to 1796. These design features represent a rejection of creeping standardisation as described by Bale (1989), projecting unique elements of Korean culture to local and global audiences.

Sport Tourism in the Natural Landscape

Natural features are central to sport tourism experiences in peripheral areas. They also present a distinct range of management issues. 'Sports like hang gliding create pressure on rural hill and scarp country, surfing on beach areas, skiing has placed pressure on mountain regions and water sports compete with one another for precious room on the limited amount of suitable inland water space' (Bale, 1989: 163). These landscapes can be quite fragile and sensitive to disturbance (Hall & Page, 2014). Sport activities on these landscapes, therefore, need to be managed in order to mitigate negative impacts (Hinch & Higham, 2004). Equally, it may be that extreme nature-based sports may foster feelings of connectedness with nature that give rise to strong desires to care for nature, and be environmentally sustainable in practice Brymer (2009).

The challenge of sustainable sport tourism development in natural areas arises for a variety of reasons. One reason is the dynamic nature of sport as reflected by the speed with which new sports are developed and diffused. The transition from an emerging sport pursued by relatively few, to a mass participation phenomenon may take place in a short space of time (Standeven & De Knop, 1999). Mountain biking, which emerged and rapidly evolved into various sub-disciplines (Hagen & Boyes, 2016; Moularde & Weaver, 2016), is a case in point. Other sports that have demonstrated a rapid rise in popularity include snowboarding, scuba diving, windsurfing, triathlon and paragliding. This dynamic presents fascinating development opportunities at tourist destinations but it also requires proactive action to establish and implement appropriate policies and management strategies to protect the natural landscapes. In some cases, the development of participation sports has taken place in the absence of a relevant legislation framework, management structure or administrative authority. Sports such as BASE jumping and bungy jumping demonstrate the challenge that management agencies may encounter with the development of new sports innovations. Extreme sports such as these often defy a single management authority (Mykletun & Vedø, 2002).

The management of the impacts of sport tourism on natural areas is a complex task. Indeed, simply measuring the impacts of tourism is fraught with difficulty. Rarely do baselines exist from which to measure change, and the impacts of tourism are challenging to disaggregate from the direct or indirect impacts of other human activities (Wall & Mathieson, 2006). The impacts of littoral sport tourism on marine flora and fauna, for example, are difficult to distinguish from those of fishing, aquaculture or the inappropriate dumping of waste materials from towns, industries, agriculture and forestry (Bellan & Bellan-Santini, 2001). The impacts of sport tourism on fragile alpine ecologies may, due to extremes of altitude and climate, require extended recovery and regeneration time frames (Flagestad & Hope, 2001). Although the visual impacts of development may be immediately apparent, more subtle changes on fragile alpine flora, growth and regeneration rates, water regulation (May, 1995) and the breeding success of rare alpine bird species (Holden, 2000) require intervention programmes and long-term monitoring.

While it is sometimes possible to identify positive and negative impacts within the social, cultural, economic and environmental contexts of a destination, these impacts are connected in a complex web of relationships. The acceptability of different impacts in combination is also viewed subjectively by different stakeholder groups, and different individuals within stakeholder groups (Wall & Mathieson, 2006). The extensive literature on sustainable tourism development and impact management is of high relevance to sport and tourism managers, who should be cognisant of the environmental impacts of sport tourism (Cantelon & Letters, 2000; Collins *et al.*, 2009; Gold & Gold, 2016).

The compatibility of sport tourism in the natural landscape

Hunter (1995) observes that sustainable tourism development requires reconciliation of human needs, as well as environmental limitations. Human needs and the benefits and costs of tourism accrue to two main groups: the hosts and the guests (Archer & Cooper, 1994). The excessive or inappropriate promotion of sport development interests over the stewardship of natural areas may give rise to congestion and crowding, social and environmental impacts or modification of the landscape in ways that are unacceptable to host communities. The sustained quality of the visitor experience must also enter into considerations of the appropriate direction and level of sport tourism development.

Sports that are pursued in the natural landscape may demonstrate varying degrees of compatibility with other sports. Incompatible motivations and goals of participation in sport may give rise to symmetric or asymmetric conflict between participants in *different* sports (Graefe *et al.*, 1984). Symmetric conflict describes a situation in which participants in two sports feel the existence of social conflict arising from the presence of the other. Jet skiers, surfers and swimmers may experience symmetric conflict giving rise in some cases to situations of physical danger. Asymmetric conflict arises when participants in one sport are adversely impacted by the presence of those engaging in a second sport, while participants in the latter may be oblivious to, or even welcome, the presence of those engaged in the former. The intrusion of technologies such as global positioning systems (GPS) and cellular telephones in nature-based sports is an increasingly common cause of social impact and conflict between sports participants (Shultis, 2000), although new technologies may become normalised over time. Sports such as orienteering and downhill skiing may be compatible with other sports, if segregated in space and/or time through appropriate management techniques. Sports that take place in coastal environments offer varying degrees of compatibility. Examples include diving, surf skiing, swimming, jet skiing, windsurfing and white-water kayaking. Consumptive (e.g. hunting) and mechanised sports (e.g. jet boat racing and water skiing) are fundamentally incompatible with other uses, as they may either irrevocably compromise alternative sporting pursuits or present physical danger to participants in other sports. This issue dictates that sports that take place in the sport tourism periphery must be carefully managed to reduce conflict.

The psychographic profile of the sport participant (Chapter 3) also determines the compatibility of different sports, and different participants in the same sport. Conflicts may arise between participants *within* a sport if the motivations of participants are incompatible. Wilderness cross-country skiing and surfing are sports that take place in environments that can be contested by numerous participants. Access to waves is managed within the surfing community by an unwritten surfing etiquette (Usher & Gomez, 2016; Wheaton, 2000, 2004). Edensor and Richards (2007) consider the increasing pressure on natural environments as leisure spaces in relation to the tensions that may arise between different groups of lifestyle sports participants, specifically skiers and snowboarders. Using a performance lens and drawing upon the embodied nature of leisure, they offer insights into how style and movement are used to create distinction and difference between groups of sport participants, which they refer to as the 'contested choreographies of the slopes' (Edensor & Richards, 2007: 97). The compatibility of sports in the natural landscape generally decreases with the seriousness of the participant or competitor (Bale, 1989).

The Impacts of Event Sport Tourism: A Paradigmatic Shift

Events hosted in central and peripheral locations offer similar opportunities to foster environmental interests and implement impact mitigation plans. The legacy of the 1992 Albertville (France) Winter Olympic Games is one of irreversible impact on the alpine environment due to intense development (May, 1995). In response, the planning of the 1994 Lillehammer Winter Olympic Games (Norway) pioneered new approaches to environmental management (Kaspar, 1998; Lesjø, 2000). The Olympic Environmental Charter (ratified in 1996) now requires that Olympic organising committees articulate and implement environmental protection policies. This charter was used for the first time in the planning of the 1998 Winter Olympics in Nagano (Japan). Cantelon and Letters (2000: 294) argue that 'it was the widespread environmental damage at the 1992 Albertville and the Savoie Region Games, and the subsequent "Green-White Games" of Lillehammer (1994), that were the historical benchmarks for the development of this policy'.

The Sydney 2000 Olympic Games represents an entrenchment, but also a significant advancement of the environmental achievements of Lillehammer 1994. Conservation, ecological restoration and the remediation of industrial sites formed an integral part of the Sydney Olympic Games development programme (OCA, 1997a, 1997b). The environmental legacies of the 1994 Winter (Lillehammer, Norway) and 2000 Summer (Sydney, Australia) Olympic Games represent a paradigmatic shift from impact mitigation to proactive environmental stewardship and habitat creation associated with event sport tourism (Chernushenko, 1996; Cowell, 1997). Environmental planning for the London 2012 Olympics included targets such as 90% of materials used in construction to be recycled, 20% of Olympic site energy requirements to be renewable and

50 miles of walking and cycle ways around the Olympic park to assist in achieving the goal of a 'car-free' event. The London Olympics were also planned in accordance with an ambitious urban regeneration legacy for east London's Stratford Lea Valley (noting that regeneration plans for the Lea Valley were in place when the London 2012 Olympic bid was won in Singapore in 2005) (Davis & Thornley, 2010). These sport events provide evidence of new perspectives on the sport tourism–environment nexus. While quantifying the environmental impacts of events (at all spatial scales) will inevitably become more important, the complexities of sports event impacts, which vary in terms of spatial scale and time, demand careful consideration of both assessment methods and organisational actions (Collins *et al.*, 2009).

Conclusion

Sport tourism environments and resources form an important part of the foundation upon which sport tourism development is built. Landscape and climate are key determinants of the attractiveness of tourist destinations. They also have considerable influence on the sport and recreational activities that tourists associate with a destination, thereby influencing destination image. The relationship between sport and the environment is a dynamic one (Standeven & DeKnop, 1999). Understanding this relationship is critical to harnessing trends that offer the opportunity for, or pose a threat to, sustainable sport and tourism development (Hinch *et al.*, 2016). Reproducible or transportable sports offer a valuable example in this respect (Bale, 1994; Weed, 2010). Transportability, while representing the threat of relocation from a sports place of origin, may also provide an opportunity to develop new or existing sports resources at specific tourism places.

This chapter also considers the sharp distinction that lies between sport tourism development in the built (central) and natural (peripheral) environments. The importance of this distinction lies in the contrasting impacts and management issues that apply to sport tourism in differing contexts, and at various scales of analysis from the local (e.g. see Pillay & Bass, 2008) to the global (e.g. see Otto & Heath, 2009). How these impacts are perceived at the local level will have a considerable bearing upon the future of sport-related tourism development (Cornelissen *et al.*, 2009; Hritz & Ross, 2010; Schulenkorf, 2009; Smith, 2010).

Part 4
Sport Tourism Development and Time

8 Sport and the Tourist Experience

The consumption of tourism and sport can embody symbols, convey meanings and provide a frame for identity making.

Moularde and Weaver, 2016: 285

Introduction

This chapter considers aspects of sport tourism in the short term and, in doing so, focuses on the tourist experience. In terms of sport tourism development and time, the tourist experience is the first and most immediate temporal element that we consider. Later, we extend the temporal dimension to consider the annual/seasonal (Chapter 9) and long-term evolutionary (Chapter 10) dynamics of sport tourism development. In this chapter, we initially consider the temporal phases of the sport tourist experience, before addressing socio-structural factors that shape and influence sport tourism experiences. The structure of the discussions that follow is provided by the fourfold classification of sport tourism presented in Chapter 2, which includes spectator events, participation events, active engagement in recreational sports and sports heritage. Informed by Morgan's (2007) experience space model, we first consider the co-creation of sport spectator and participant experiences by way of unique socio-cultural interactions that unfold through the interplay of activities and people in a sport setting that is place-situated. We then turn our attention to tourist experiences of active engagement in sports, and heritage sports, respectively. We conclude this chapter with thoughts on how visitor experiences are shaped and influenced by the sport and tourism systems at a destination.

Temporal Phases of the Sport Tourist Experience

While here we consider the immediate tourist experience, it is noteworthy given the temporal focus of Part 4: Sport Tourism Development and Time, that the tourist experience itself represents the sum of several distinct processes that unfold over time in five phases: anticipation, travel, visitor experiences at the destination, travel back and recollection (Clawson & Knetsch, 1966; Manfredo & Driver, 1983). The anticipation phase of the tourist experience requires an understanding of information search, decision-making, planning and the formulation of expectations, each of which are affected by travel motivations. Travel to the destination forms an important part of the tourist experience as it influences length of stay and the infrastructure and service needs of the tourist. Engagement in sporting activities and the feelings and behaviours of sport tourists are key dimensions of the visitor experiences; these represent the interplay of activity, people and place (Weed, 2005). These engagements vary significantly with, for example, the relative priority placed on specific aspects of the pursuit of sport, and other touristic activities at a destination (Higham & Hinch, 2009; Morgan, 2007). Visitor experiences at the actual destination, therefore, form one part of the subject of this chapter, given that they are set within the wider context of people's lives prior to and following the tourist experience (Morgan, 2007).

It is useful to briefly consider the temporal phases of the sport tourist experience. The typology of sports tourists presented in Table 8.1 serves as a starting point from which to consider the pre-trip phase of the sport tourist experience. Different sport tourist types are likely to vary considerably in terms of information search, decision-making and planning. Anticipation of spectator events, for example, may be situated within wider team fandoms and the search for collective experiences in association with team support and identity building (Jones, 2000), although Stewart (2001) highlights the diversity that exists within the broad terminology of the sports fan (Laverie & Arnett, 2000). Avenues of information search in the pre-trip phase will stand in significant contrast between sport tourist types and between first time and repeat visitors (Taks *et al.*, 2009). Those

Table 8.1 Sport tourist typologies

Sport tourist typology (Glyptis, 1982)	Parallels with subsequent typologies
General holidays with sport content	Incidental (Jackson & Reeves, 1997) Incidental (Reeves, 2000)
Specialist or general sport holidays	Sport activity holidays (Standeven & De Knop, 1999) Independent sport holidays (Standeven & De Knop, 1999) Sporadic (Reeves, 2000) Occasional/regular (Jackson & Reeves, 1997) Occasional sports (wo)men (Maier & Weber, 1993) Mass sports (Maier & Weber, 1993)
Upmarket sport holidays	Organised holiday sports (Standeven & De Knop, 1999) Occasional/regular (Jackson & Reeves, 1997)
Elite training	Top performance athletes (Maier & Weber, 1993) Dedicated/driven (Jackson & Reeves, 1997)
Spectator events	Passive sports on holiday (Standeven & De Knop, 1999), including casual and connoisseur observers Passive sports tourists (Maier & Weber, 1993)

who pursue specialist or general sport holidays are perhaps most likely to respond to attraction markers, both contiguous and detached (Chapter 2), to establish or raise awareness of a destination. Those who engage in elite training and competition, by contrast, operate under unique circumstances in terms of planning and anticipation of the experiences that they set out to achieve at the tourist destination, and are likely to operate in compliance with the directives of professional sports organisations (PSOs) (Higham & Hinch, 2010).

Socio-Structural Factors Shaping Sport Tourism Experiences

Visitor expectations may vary significantly among participants both between different sports and within a given sport (Chapter 3). Expectations and desired experiences are a function of the lifestyles, attitudes and personalities of individual sport tourists and are likely to vary considerably with demographic profile, life and travel career stage, sport and personal experience or involvement in a sport or sports team (Gibson, 2005; Pearce, 1988; Schreyer et al., 1984; Watson & Roggenbuck, 1991).

Gibson (2005: 59) provides particularly useful insights into important aspects of the sport tourism experience, highlighting that 'findings from a range of studies in leisure, sport and tourism... have concluded that the relationship between needs and activity choice is quite complex'. Her systematic review of the sport tourism experience emphasises that motivations and behaviours are multidimensional. Behaviour choices vary between individuals, and evolve over time, based on the need for stability and variety (or arousal). Those who seek optimal levels of stimulation seek experiences that are adventurous, challenging and novel, whereas others may seek experiences that are personally enriching but secure and predictable.

Research into tourist motivations is concerned with why people travel, the benefits that they seek and the experiences that they pursue to satisfy their needs and desires (Cooper et al., 1993). Tourist motivation is a function of the self-perceived needs of the traveller, which drive the decision-making process and the purchase of tourism products (Collier, 1989). The motivational profile of the traveller is a combination of intrinsic and extrinsic factors. These factors have been described in terms of push (psychological) and pull (cultural) factors (Dann, 1981); the need being to match motivations (push factors) with destination attributes (pull factors) (Crompton, 1979; Gibson, 2005). In the context of sport specifically, push factors may extend to the desire to achieve sport career or serious leisure objectives that cannot be achieved at home. Green and Chalip (1998), for example, report that sports people may be motivated by competency and mastery, which can be achieved in various levels of competition, but also by affiliation and socialisation needs, and to develop or embellish subcultural identity, which may be best served in a collective tourism context.

Pull factors include price, destination image and marketing and promotion. Destination image, which is a function of physical and abstract attributes (Echtner & Ritchie, 1993), plays an important part in the formulation of expectations. Physical attributes include attractions, activities, sporting facilities and physical landscapes. Abstract attributes are less readily measured, and include atmosphere, crowding, safety and ambience. Again, specifically in terms of sport experiences, pull factors may relate to the search for competition or the achievement of sporting experiences that are unique to particular places (Hinch, 2006; Hinch & Higham, 2005). Unique experiences are usually bound in both space and time (e.g. sports events, sports seasons). These pull factors may also influence the perceived needs of tourists in the anticipation phase of the travel experience.

Sport tourism, like other forms of travel, entails a set of motivations that are established in anticipation of the fulfilment of desired needs. Stewart (2001) identifies a range of factors that motivate fans to travel to support their teams and these fall within the gambit of push and pull factors. Push factors include release from everyday life, the search for camaraderie, to develop friendships and a sense of belonging and the opportunity to do things that cannot be done at home (e.g. to enhance one's standing within a sport subculture). In contrast, pull factors that may motivate event sport tourism include the unique interplay of a significant sports activity and people and place that distinctive tourism destinations offer (Weed, 2007). Pull factors may extend to the significance of a spatially and temporally bound context and the state of flow that may be associated with uncertainty of outcome and post-competition revelry and celebration. Tourist motivations are critical to understanding why people do or do not travel, their choice of destination and other aspects of tourist behaviour.

Spectator Event Experiences

The complexity of the sport tourism experience is such that it is necessary to address the motivations and experiences of sport tourists within the context of different forms of engagement in sport and tourism. This approach will serve our purposes in this chapter, although later in the chapter we will consider that ways in which the defining lines between different forms of sport tourism have become increasingly blurred and indistinct. Event experiences are sought out by sport spectators who travel to experience elite athletic performances, or who follow a sports team as individual fans or as part of a wider sports fandom (Osborne & Coombs, 2013).

Morgan (2007) adopts a sociological perspective on the management of spectator event experiences. In addressing the interplay of event spectators and the place of competition, he highlights two perspectives on event spectator experiences:

1. Managerial view: An experience as a type of product or service to provide an added-value offering.
2. Consumer behaviour view: An experience as having emotional, symbolic and transformation significance for the individual involved. (Morgan, 2007: 362)

The managerial approach describes a commodified product that is developed to meet the needs of sport tourists as a homogeneous group. The consumer behaviour approach, as described by Morgan (2007: 361), views the sport tourist experience as a 'subjective emotional journey full of personal, social and cultural meanings'. Morgan's (2007) critique argues that the commodification of sport experiences through the provision of predefined, controlled and stage-managed experiences may reduce the live experience of spectator sport to the equivalent of a television viewing or live streaming experience, in which the spectator is an observer; passive in all aspects of the unfolding event. Indeed, the quality of the live viewing experience may in some respects be inferior to the electronic media or virtual experience. Alternatively, by elevating the subjective and emotional elements of the experience as an interplay of a unique sporting occasion taking place in a local place-situated context that is central to the occasion and in association with the interactions of travelling supporters and local hosts (Weed, 2005) renders the sport spectator experience unique (Morgan, 2007). Interactive online media represent the new middle ground of live sport experiences.

In illustrating a sociological perspective of the visitor experience, Morgan (2007: 363) presents an experience space model (Figure 8.1) to explain that 'the experience is created by the interaction between the activities and the places provided by the destination and the internal motivations and meanings brought by the visitors'. Figure 8.1 illustrates the motivational pull factors, which centre on the physical attributes and image of the destination, as mediated by destination management and marketing activities. This represents the *in situ* destination setting where the sport experience takes place. The model then outlines the motivations, meanings and identities that tourists bring to the place of competition. In the case of sport tourism, the spectator experience will be heavily influenced by personal meanings and the collective social identities of communities of supporters (Morgan, 2007). The experience space allows for the co-creation of experiences based on social and cultural interactions that represent the interplay of activities and people in a sport setting that is place-situated.

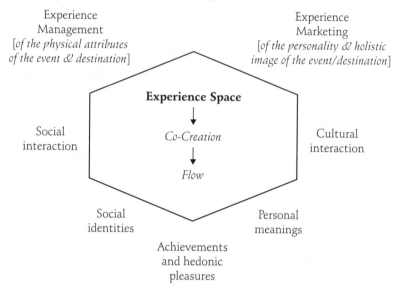

Figure 8.1 The experience space (Source: Morgan, 2007: 362)

Morgan (2007) highlights three internal elements of the spectator experience. The first relates to the personal benefits of *hedonistic enjoyment and achievement*. Novelty, impulse, entertainment and surprise are essential elements of experience that can be achieved in the tourist space. So too, according to Morgan (2007), is total absorption in an activity, which can give rise to the elevated state of 'flow' (Csikszentmihalyi, 1992) (see Figure 8.1). The second is *social interactions*, which lie at the heart of the intense collective experiences of spectator sports. These interactions provide a setting in which individual and social identities are forged, and where social and environmental circumstances may give rise to a sense of group identity or 'communitas' (Chalip, 2006; Hinch & Higham, 2004; Morgan, 2007; Weed, 2005). Interactions with local people, both in terms of integration (shared experiences) and differentiation (development of social identity), may be critical elements of the sport tourism experience (Weed & Bull, 2004). The third key element is *meaning and values*. The meanings and values associated with a sport spectator experience are derived from the cultural background of the visitor and his or her subjective understanding and interpretation of the historical and cultural context of the sport and place where it takes place. 'Sporting allegiances and cultures are therefore deep-set and subtle expressions of personal identity' (Morgan, 2007: 368).

Thus, in terms of sport tourism, the significance of the place of competition (the destination) rests with the social and cultural interactions that occur when people travel, tour and gather at places of competition and engage in the tourist experience (Weed, 2007). The social experience is place-situated and co-created through the interactions of hosts and guests. Within this context, hedonistic enjoyment and achievement, social interaction and meaning and values lie at the heart of the sport tourist experience (Morgan, 2007). This point corroborates the findings of studies of sports fans, which demonstrate the importance of the social experience in terms of travelling for sport spectatorship (Fairley, 2003).

Morgan's (2007) work provides intriguing insights into the centrality of personal and subjective sports histories to the performance of sport-related event experiences of activities, people and place. This, he argues, gives emphasis to the importance of allowing visitors to 'co-create' their experiences, rather than to manage and impose event experiences. National and/or regional tourism organisations, as well as sports organisations and event managers, need to better understand the expectations and desired experiences as having 'emotional, symbolic and transformation significance for the individual involved' (Morgan, 2007: 362). This raises fascinating questions regarding how sport event hosting agencies, with tourism destination managers, collaborate to produce events that allow for the co-creation of sport tourism experiences (see Focus Point 8.1).

Focus Point 8.1: Co-Creating Event Experiences – Japan 2019 Rugby World Cup

James Higham, Tom Hinch and Adam Doering

Successive Rugby World Cups (RWC) have served to make or remake important elements of national identity. Perhaps the most vivid case is that of the 1995 RWC, which was hosted in the post-Apartheid Republic of South Africa (Steenveld & Strelitz, 1998). In a national sport historically characterised by Afrikaner (white) nationalism (Grundlingh, 1994; Laidlaw, 1999), the 1995 RWC played a critical role in forging a post-Apartheid national identity based on the branding of the 'Rainbow Nation' (Maingard, 1997; Nauright, 1997a, 1997b). This contributed to the 'remaking' of South Africa and the emergence of South Africa as a global tourism destination (Picard & Robinson, 2006). The prominence of the contributions that RWC has made to national identity, cultural politics and nation (re)making and (re)imaging raises interesting questions as to the cultural politics of Japan's hosting of RWC 2019. Morgan's (2007) theorisation of event delivery and host–guest experiences (Figure 8.1) may usefully inform this question, given

(Continued)

Focus Point 8.1: (Continued)

the attention he pays to the interplay of external (as interpreted by the event organiser) and internal (as perceived, understood and interpreted by event attendees) elements. External elements relate to the image and brand of the destination in terms of physical destination attributes and the holistic image of the destination and the event. These elements, as constructed through event planning, marketing and management, '... make it possible for experiences to happen, but they do not actually create them' (Morgan, 2007: 362); rather they provide a context within which unique tourist experience can occur. Internal factors are those that motivate and engage tourists to attend an event. These include the personal meanings and social identities that arise from the engagement of tourists in sport, which are critical to both event production and the visitor experience. To understand the importance of internal factors, it '...is necessary to uncover the personal meanings which give the event significance in the lives of individuals' (Morgan, 2007: 362).

Japanese rugby represents an alternative multi-ethnic Japan, but RWC 2019 organisers are likely to be incentivised to promote the opposite, due to the conditions of having to create a cultural niche in a global marketplace. This may serve to inadvertently reinforce the same primordial national identity through sport tourism that leads to the exclusion of difference and cultural diversity in Japan. Tourism is known to accentuate essentialised national identities rather than seeing the complexity in them. In sport, traditional 'patriotic games', particularly major sports events, have tended to do the same. However, mobility and globalisation are increasingly impacting the nexus of sport, tourism and identity. Sport labour mobilities in particular have contributed to increasing diversity and have at the very least raised questions about traditional understandings of identity. This is very much the case in Japan where the cultural diversity of the national men's rugby team, which for many years has included players born in New Zealand, Samoa and Tonga (among others), is challenging the entrenched view of ethnic homogeneity and mono-culturalism. Tensions arise when, despite these increasing diversities, events like the RWC are used for conservative rationalities and legitimisations. Beyond the field of competition, one of the more intriguing aspects of the 2019 RWC will be the way the politics of identity play out.

It is interesting to consider cultural politics in terms of the elements and interpretations of national identity that will be portrayed in the delivery of the Japan 2019 RWC. It is also important to consider how RWC 2019 event organisers will deliver elements of experience management and experience marketing (Figure 8.1) in relation to (contested) aspects of national identity. This, in turn, raises questions about how RWC 2019 organisers will negotiate the longstanding tensions of Japanese culture associated with the national rugby team (Light *et al.*, 2008). The cultural diversity of the Japanese national rugby team is such that the 2019 World Cup will offer an opportunity for Japanese national identity to be reimagined. Identity politics will shape the interplay of event organisers and the event tourists who will jointly co-create the event experience.

Key Reading

Hinch, T.D., Higham, J.E.S. and Doering, A. (2018) Sport, tourism and identity: Japan, rugby union and the transcultural maul. In C. Acton and D. Hassan (eds) *Sport and Contested Identities: Contemporary Issues and Debates* (pp. 191–206). London/New York: Routledge.

Light, R., Hirai, H. and Ebishima, H. (2008) Tradition, identity professionalism and tensions in Japanese rugby. In G. Ryan (ed.) *The Changing Faces of Rugby: The Union Game and Professionalism Since 1995* (pp. 147–164). Newcastle: Cambridge.

The phenomenon of travelling to attend and experience sport extends to those who engage in one-off or recurring mobilities, which vary in terms of sport experiences between major sports events, as discussed above, and the enduring fandoms associated with championship teams (Higham & Hinch, 2010). Sports fandoms offer interesting insights into aspects of sport and tourist experiences, given the diversity of factors that motivate sports fans to travel in support of their team. Stewart (2001) offers a typology of Australian team sport watchers, which demonstrates the diversity that exists among sports fans. The motivations held by each of these fan categories shape and influence visitor experiences at the host destination.

(1) *Passionate partisans:* Hardcore supporters who attend games regularly, regardless of inconveniences; their moods and identities are closely linked to the successes and failures of their team.
(2) *Champion followers:* Less fanatical, and change their allegiance or their allegiance remains held in abeyance until their team starts winning some games.
(3) *Reclusive partisans:* Interest in the game and commitment to the team is strong, but they attend games infrequently. Interested in the team more so than the game.
(4) *Theatregoers:* Primarily seek entertainment through sport but are not necessarily attached to a particular team.
(5) *Aficionados:* Attracted to exciting games, and also to games that involve star players. Interested in the demonstration of skill, tactical complexity and aesthetic pleasure, which take priority over the outcome of the game.

No doubt, most fans are inspired to travel to watch their teams succeed and to experience the euphoria of championship success. For many, too, the sport experience extends to the excitement of the carnival atmosphere associated with a moment in time when intense sport competition occurs. Manifestations of national carnivalesque may emerge in these settings (Giulianotti, 1996). Sport then becomes a cultural experience, not only in terms of sport as a manifestation of culture (Bale, 1989) but also the collective behaviour of fans from different regions, countries and continents (Giulianotti, 1995a, 1995b; Morgan, 2007). These expressions of individual and collective identity coalesce within the context of the destination where competition takes place. So, spectators who experience an important sporting occasion at a given time are also subject to the cultural experience of a specific place. Indeed, for some sports fans, the experiences of place, perhaps due to historical associations or through visiting tourist attractions or iconic local places and historical sites, may be critical aspects of the sport tourist experience (see Focus Point 8.2). This highlights sport tourism as the integration of activity (sport), people (sports fans) and place (the place of competition) (Weed & Bull, 2004). The uncertainties of sports performances, circumstances and outcomes, combined with the unique atmosphere of world championship events, may function as very powerful tourist attractions (Hinch & Higham, 2004).

Stewart's typology demonstrates that sports fans cannot be considered a homogeneous travel market due to the variation in drivers motivating fans to travel in support of a team (Borland & MacDonald, 2003). While performance and outcome may be important to many fans, it is also the case that social identity can be constructed and reinforced through fandom membership whereby 'sport becomes a pivotal means of signifying loyalty and commitment, producing enduring leisure behaviour' (Jones, 2000: 285). Identification with a sport subculture may be an important motivation for both participants and spectators. Green (2001: 5) notes that 'interactions with others are at the core of the socialisation process and provide avenues through which values and beliefs come to be shared and expressed'. Sport tourism may, therefore, be motivated by a celebration of subculture through participation, spectatorship, association or interaction (e.g. through related non-sport activities at the location where the sport or competition takes place) (Green & Chalip, 1998).

Focus Point 8.2: Sport Event Experiences that are Bounded in Space and Time

Canada's victory over the United States in the Vancouver 2010 Winter Olympics men's ice hockey final triggered scenes of jubilation among

Canadians in Vancouver (and elsewhere around the world). The spontaneous celebrations that occurred in Vancouver were spatially and temporally bounded – it was *the place to be* at that moment in time. Similarly, the 1998 FIFA World Cup was hosted, and the final won in front of the home crowd, by France. The path that the French team negotiated to the final was a tenuous one. Their quarter-final against Paraguay was won in extra time by a 'golden goal'. In the semi-final, France was victorious over Italy, a three-time champion, in a game that was level after full time and extra time, and decided in a gripping penalty shootout. In the final, on 12 July 1998, the French team administered the *coup de grace*, in a complete performance 'defeating the ultimate adversary, the football nation of legend, Brazil' (Dauncey & Hare, 2000: 344) by the convincing margin of 4–1. 'On the night of 12 July ... there was an outpouring of joy and sentiment that was unprecedented since the Liberation of 1944... Huge numbers of people poured onto the streets in spontaneous celebration. In Paris, hundreds of thousands gathered again on the Champs Elysées the next day to see the Cup paraded in an open-topped bus. For all, the victory was an unforgettable experience' (Dauncey & Hare, 2000: 331). A global television audience, at the time the largest television audience for a single sport, watched these events unfold. There can be little doubt about the emotion associated with defining sporting moments that are experienced only at a given place in a given moment of time. Laidlaw (2010: 51), in reference to sport and identity, describes some sports as a 'barometer of feelgood', noting that 'nobody who was around when... South Africa won the (rugby) World Cup in 1995 would be under any illusion as to the collective high that those moments brought to a whole nation'. Giulianotti (1996: 323) notes the significance of the US 1994 FIFA World Cup in terms of the Irish team fans being based in New York and Orlando, during which 'the soccer culture helped to promote a fresh sense of Irish identity, as beyond nation-state boundaries or territorial claims'. These celebrations were characterised by an absence of football hooliganism, interactions of Irish fans with other supporter cultures and expressions of carnival fandom in places of historic significance to the Irish diaspora.

Key Reading

Dauncey, H. and Hare, G. (2000) World Cup France '98: Metaphors, meanings and values. *International Review for the Sociology of Sport* 35 (3), 331–347.

Giulianotti, R. (1996) Back to the future: An ethnography of Ireland's football fans at the 1994 World Cup Finals in the USA. *International Review for the Sociology of Sport* 31 (3), 323–347.

Participation Event Experiences

The experiences of event participants are most logically considered in relation to the level of competition, which varies from professional/elite to the amateur ranks, which in turn varies from serious amateur athletes to recreational and social sport participants. The experiences of each could not be more contrasting (Higham & Hinch, 2010). Professional and elite athletes face unique travel circumstances associated with achieving the right physical and emotional states required to attain the highest levels of competitive performance. Professional athletes are the business travellers of sport tourism (Hodge *et al.*, 2010), and often travel under the directives of PSOs and/or team management. Woodman and Hardy (2001) highlight a range of causes of organisational or work-related social psychological stress in elite sport, many of which bear upon or relate directly to the circumstances that athletes face at places of competition. They consider organisational stress caused by environmental, personal, leadership and team factors, highlighting the need for consistency, predictability, support and, in some cases, control over the social and physical environment prior to and during competition. While many causes of stress among elite sport participants relate to the sport itself (e.g. loss of form, injury, selection and financial hardship), stress may also arise from specific destination factors (unfamiliarity, homesickness, loneliness, poor accommodation, inconvenient or substandard training facilities) (e.g. disabled athletes, see Darcy, 2003).

It is noteworthy, however, that other causes of organisational stress highlighted by Woodman and Hardy (2001) may be mitigated or eliminated with careful consideration given to locational and living arrangements when athletes are based in foreign countries while preparing for and engaging in competition. Accommodation and nutritional arrangements can be closely managed at places of training and/or competition. Boredom and isolation in the training environment may be mitigated or eliminated through strategies such as integration into local communities of competitors, as well as through the tourist experiences achieved in novel and stimulating places in an attempt to establish a balance between the business of sport and escape from the pressures of competition. It is also likely that team culture may be enhanced through the shared experiences of new, unique or iconic places.

For elite athletes and professional sports team members, optimum competitive performance is a function of creating the right environment for preparation and competition (Hodge *et al.*, 2001). Sports managers make reference to 'team culture' as an essential element of a successful team. In seeking to establish a culture for team success, team management will consider all aspects of the team environment that may influence competitive team performance. Travelling athletes face the additional challenges associated with unfamiliar living and playing conditions, compromised daily routines and physiological stresses caused by trans-meridian travel (Higham & Hinch, 2009). 'To counter the perceived disadvantages of away games, professional sports teams have invested heavily in addressing these factors in their quest to win away games' (Francis & Murphy, 2004: 78). A major focus of this effort has been to create the desired team environment, which may be achieved by creating familiar and secure environments for competitors at the place of competition.

Achieving a strategic fit between the destination where teams/athletes are based and team culture is an important strategy. Creating the right environment requires the achievement of a balance between familiarity and novelty. Many sports teams seek an 'environmental bubble' in which to prepare for competition in a setting that is low in stress and high in familiarity (Higham & Hinch, 2010). However, insular, familiar and routined environments can become suffocating for athletes. Therefore, creating the right environment may extend to locating the team at a destination that offers opportunities to engage in tourist and leisure activities that can serve the function of bringing the desired level of novelty to the competition preparation phase. Distinctive and interesting tourist and heritage attractions may serve as a release from the pressures of competition, and may also provide mental focus for players. Thus, the optimum experience for elite athletes at destinations where competition takes place involves 'tour balance': the striking of an appropriate balance between athletic preparation and engagement with the place of competition (Hodge et al., 2008).

Distinct from elite athletes are those who engage in sport tourism as active participants, at various levels of competiveness. The Masters' Games phenomenon embodies much of this diversity in one single event (Ryan & Lockyer, 2002). Ryan and Trauer (2005: 183) identify four types of competitors who may engage in Masters Games competition. They include:

(1) *The Games Enthusiast:* High levels of involvement in his or her sport, motivated both by the intrinsic awards of physical well-being and the social interactions that come with participation.
(2) *The Serious Competitor:* Motivated by successful competition and high involvement.
(3) *The Novice/Dabbler:* Engages in sport for reasons of physical fitness, and for reasons of participation rather than competition.
(4) *The Spectator:* Knowledgeable and interested in sport but exhibits minimal direct engagement.

The diverse motivations and broad catchment of participants in Masters sport events explain the rapid growth in Masters competitions, from regional and national to world championships, which now attract

Case Study 8.1: (Continued)

competing priorities and the impact on partners and other family members (Kennelly *et al.*, 2015).

Consequently, in a recent study we have shifted focus to exploring the impacts of serious leisure on amateur athletes' spouses (Lamont *et al.*, 2015). Previous studies have identified three models of spousal leisure interactions (Gillespie *et al.*, 2002; Hultsman, 2012). The first encapsulates circumstances where both spouses share an interest and participate in a leisure activity together. The second model is where one spouse is highly committed to an activity and the other spouse is not committed but provides support to the committed spouse. In the third model, one spouse is engaged in serious leisure and competes in events related to the activity; this spouse is therefore positioned as the 'dominant' competitor, with the other spouse attending events to enact a support role. Our data suggested an additional 'dominant participant/secondary participant' model reflecting situations where one spouse within the couple was the dominant participant in endurance sports, while the other spouse was a secondary (less involved) participant. Although both spouses shared an interest in the activity, the secondary participant's involvement was constrained by enacting a support role to the dominant participant.

Indeed, our data indicated that for some couples, it is extremely difficult for both spouses to be highly committed to leisure activities, particularly when the couple have young children. Thus, our research found that one spouse within a couple tended to adopt a support role for their athlete spouse, spanning both day-to-day life and event travel contexts. 'Supporting' spouses tended to report enduring frequent absenteeism by their athlete partner which evoked a series of interconnected stressors such as increased domestic responsibilities and anxiety over their athlete's safety and well-being. Stressors, however, were counteracted by an array of perceived benefits such as enhanced family togetherness and travel opportunities (Kennelly *et al.*, 2015). Further research is required on the paradoxical cycles in which athletes and their spouses tolerate significant stressors offset by the benefits that sustain event travel careers in the longer term. From a supply perspective, more research is needed on developing event portfolios with the capacity to build sustainable sports tourism destinations for both athletes and their all-important, but often neglected 'non-participating entourage'.

Key Reading

Kennelly, M., Moyle, B.D. and Lamont, M. (2013) Constraint negotiation in serious leisure: A study of amateur triathletes. *Journal of Leisure Research* 45 (4), 466–484.

Lamont, M., Kennelly, M. and Wilson, E. (2012) Competing priorities as constraints in event travel careers. *Tourism Management* 33 (5), 1068–1079.

Additional references cited in this case study are included in the reference list.

Active Engagement in Recreational Sports

Active engagement in recreational sports is an aspect of sport tourism that offers fertile ground for critical research. Hagen and Boyes (2016), for example, offer in-depth insights into the experiences of active mountain biking participants. They explore the relationship between mountain bike subculture and riding practice in relation to heightened and deep affective experiences. Theoretically informed by Bourdieu's theories of field, habitus and capital, they explore aspects of social history, race governance and forms (cross-country, downhill and enduro), and the forces of mainstreaming, fragmentation and resistance, media consumption and commercialisation, and gender. These aspects of a participation sport have important implications for participant experiences of sport in terms of the socially constructed nature of the mountain bike subculture and affective sensations (see Focus Point 8.3).

Focus Point 8.3: Mountain Biking Affect

Scarlett Hagen and Mike Boyes

Gravity mountain biking is a sport that provides participants with a range of destinations, landscapes, terrain and track obstacles. Locations such as Queenstown (New Zealand), Whistler (Canada) and Morzine (France) are widely recognised as *the* summer destinations for mountain bike enthusiasts. Trail managers in each location have built trails to showcase the natural geological features of these mountain destinations. Important factors for well-designed trails include site selection, terrain, gradient, elevation and the inclusion of specific obstacles. The trails need to descend the hillside in unison with the natural landscape, and to work with the terrain to be predicable for riders at high speeds. Queenstown provides rocky terrain on steep slopes, whereas Whistler has longer trails with a large number of jump-trails. On such trails, riders travel over multiple jumps, aiming to get their level of desired air-time. The terrain in Morzine is termed 'hero-dirt' as the tracks are smoother and flow down the hillside with ease. On these trails, rocks are limited and the trails are wider and considerably safer for beginner riders. Each destination has a unique point of difference.

(Continued)

Focus Point 8.3: (Continued)

The technical challenges in these locations present riders with opportunities for affective experiences and self-expression. Affective terrain is considered to include obstacles such as jumps, drops, berms, chutes and fast sections that combine to create distinctive ride moments. As riders travel over these sections of track, there is the opportunity to experience weightlessness; gravity acting *in* and *on* their body, and physical craving for faster speeds. As riders' travel along mountain bike tracks, they experience the interlinking interaction between the terrain, their bike and body. Each person's level of riding ability determines the depth of his or her affective experience. The more skilled a rider, the more opportunities there are for a wider range of mountain bike experiences. The strongest ride affects are evoked by the accumulation of affective moments on various technical and challenging obstacles. Riders use the term 'flow' to refer to the way they ride down the trail smoothly, maintaining their speed without interruption. As riders become more technically advanced, they achieve more consistent levels of flow. Riders with high levels of ability are able to maintain flow on trails that are rough, bumpy and highly technical. An intimate understanding of mountain bike riding practice in relation to affective experiences is critical to the competitiveness of tourism destinations that target dedicated mountain bikers and those who engage in mountain biking as a form of serious leisure.

Key Reading

Hagen, S. and Boyes, M. (2016) Affective ride experiences on mountain bike terrain. *Journal of Outdoor Recreation and Tourism* 15, 89–98.

A key point to emerge from research into active participation in sports is the importance of social identity and belonging. Whether active engagement in sport takes the form of serious leisure or recreational participation, the fact that sport and tourism offer a frame for identity making has grown in importance in recent years (Higham & Hinch, 2010). Moularde and Weaver (2016: 285) note that 'the consumption of tourism and sport can embody symbols, convey meanings and provide a frame for identity making'. Identity construction has provided important insights into participation in lifestyle sports such as surfing, windsurfing and mountain biking (Hagen & Boyes, 2016; Wheaton, 2004). Humphreys (2011) deploys Bourdieu's concepts of social and cultural capital to examine how golfers understand the reputation of certain golf courses, bestowing them with elevated standing which, in turn, reflects upon their own status if they visit and play on those courses. In other sports, consumer identification with a sport subculture may be a critical element of the visitor experience (see Chapter 6). Green (2001: 5) notes that 'interactions with others are at the core of the socialisation process and provide avenues through which values and beliefs come to be shared and expressed'. Sport tourism may, therefore, be motivated by a celebration of subculture through participation (Green & Chalip, 1998).

Active participation in sports, therefore, represents a diversity of forms of engagement, which is reflected in an equally diverse range of desired tourist experiences. The sport of golf demonstrates the varied range of tourist motivations held by sport participants (Hudson & Hudson, 2010). The motivations underpinning travel to play golf, and as a consequence, the sport tourism experiences of golfers, differ between distinct sport tourist types (Table 8.2). These cannot be generalised directly to other sports due to the unique rules, competition structures and elements of play that characterise each sport (Chapter 2). However, an understanding of the motivations that sport tourists hold towards sports such as golf is critical to fostering desired visitor experiences (Humphreys, 2011). Humphreys and Weed (2014) present the findings of a study that examines aspects of negotiation and compromise, as influenced by lifestage, on decisions to engage in golf tourism and to play golf during family trips to a destination. Drawing on two tiers of the decision-making unit (DMU), Humphreys and Weed (2014) provide insights of importance to destination managers. Similar studies that examine the tourist motivations and decision-making processes associated with other sports are certainly well justified.

Table 8.2 Sport motivation profiles of sport tourists who play golf

Sport tourist types (Glyptis, 1982)	Primary motivations	Destination attributes	Secondary activities
General holidays with sports content	Various business or leisure travel motivations	Vary with primary motivations (existence of a golf course is incidental)	Playing golf, among other things
Specialist sport holidays	Pilgrimage to the heartland of golf. Emulating icon players	Grand slam and other championship courses	Nostalgia sport tourism
General sport holidays	Golf as one part of a suite of visitor activities	Single integrated resorts	Family-based activities
Upmarket sports holidays	Golf as a specialised visitor activity	High degree of luxury. Second home developments adjacent to golf courses	Domestic and social activities
Elite training	Seek competition and being challenged by a range of golf courses	Networks of golf courses forming golf regions	Coaching clinics, professional advice, purchase of equipment

Source: Adapted from Glyptis (1982) and Priestley (1995).

Heritage Sport Experiences

Heritage and nostalgia have emerged as an important avenue of personal and social identity and a key element of authentic sport experiences (Higham & Hinch, 2010). Fairley and Gammon (2005) address sport and nostalgia in terms of artefact and experience. Heritage sport experiences have developed in ways that offer parallels with the study of heritage tourism (Gammon & Ramshaw, 2013). Heritage sport tourism, as originally conceived by Redmond (1990) in the North America context, centred on the sport place or artefact, as expressed in terms of tourist visits to sport museums, halls of fame, themed bars and restaurants. While some sports facilities have developed their own heritage values over time, Ramshaw (2011) observes growing interest in the development of new sports facilities with strong heritage elements. The phenomenon of the 'retro park' emerged in the early 1990s, as a means of reclaiming the strong heritage values that can enrich the sport experience (Higham & Hinch, 2010; Ramshaw, 2006).

Initially, then, heritage sport experiences were derived from built structures; either stadiums and other sports venues that may develop their own aura of history over time (Bale, 1989) or (re)constructed parks, stadiums, museums and halls of fame (Ramshaw, 2011). While the commercialisation of sports heritage matured in these forms in North America (Gammon, 2002), the growth in demand for heritage sport tourism has resulted in the development of heritage experiences that are diverse and multifaceted, which is now reflected in the academic literature (Fairley & Gammon, 2005). Heritage has been a critical element of many sport tourism experiences. Fairley (2003) notes that nostalgia is a driver of group repeat travel to experience sports, playing an important part in the recruitment and socialisation of new group members. Research that explores the means and ways of building heritage values and embellishing nostalgic experiences of sport is of considerable interest. Derom and Ramshaw (2016) offer insights into the potential for managers of participation sports events to leverage the heritage values of their sport not only to promote participation, but also to foster tourist engagement with the wider heritage of a tourism destination (see Focus Point 8.4).

Focus Point 8.4: Tour of Flanders Cyclo Event

Inge Derom and Gregory Ramshaw

Major sport events, including well-known cycling races such as the Tour of Flanders, may be leveraged to attract international tourists and potential investors to the host community and region. Event leverage describes the strategic process of aligning event and destination-related resources with newly developed initiatives to create additional benefits for the host community (Derom & VanWynsberghe, 2015). In an effort to attract more tourists to the region of Flanders, cities have been transformed by the major cycling race, for example by the construction of the permanent Tour of Flanders Museum in the arrival host city of Oudenaarde. The Tour of Flanders is organised annually in conjunction with the Tour of Flanders Cyclo, a heritage-based active sport tourism event that is open to the general public and attracts 16,000 participants from across the world. To examine opportunities for event leverage associated with active sport tourism events, data were collected from event participants prior to the 2013 edition of the Tour of Flanders Cyclo. An online survey was completed by a sample of national ($n = 650$) and international ($n = 441$) event participants. Comparing the two samples of event participants within the framework of event leverage has shown that different event and destination-related resources are important for different groups of event participants, who are consequently in search of different event experiences. In the case of the Tour of Flanders Cyclo, which makes explicit associations to Flanders' cycling heritage, sport heritage and more specifically 'active' sport heritage has been identified as an important resource to promote participation among active sport tourists. This is particularly important for international event participants who are more committed to experiencing and pursuing the event's heritage, as active participants as well as passive spectators, when compared to national event participants. This motivation to experience 'active' sport heritage demonstrates that event managers may wish to permanently embed heritage (via the route, history and atmosphere of the event) within the host community and region to highlight a specific heritage that separates the event from competing events and to attract international tourists beyond the duration of the event.

Key Reading

Derom, I. and VanWynsberghe, R. (2015) Extending the benefits of leveraging cycling events: Evidence from the Tour of Flanders. *European Sport Management Quarterly* 15 (1), 111–131.

Derom, I. and Ramshaw, G. (2016) Leveraging sport heritage to promote tourism destinations: The case of the Tour of Flanders Cyclo event. *Journal of Sport & Tourism* 20 (3/4), 263–283.

Sport and Tourism Systems: Visitor Experiences

The sport tourist experience can be described and studied in various ways but fundamentally it represents a combination of tangible (physical attributes) and intangible (emotions and feelings) elements (Weed, 2010). The experiential approach involves understanding the emotions and feelings that comprise the visitor experience. These emotions may include joy, relief, exhaustion, euphoria and dejection, which arise from victory, defeat, camaraderie and sense of history, which may arise through event and competition spectatorship and/or active participation in sports, or be relived through nostalgic experiences of sports. Visitor experiences are a function of the motivations and desired experiences of the tourist. Walker *et al.* (2010), for example, use 'activity mode' and 'task orientation' to explore the experiences of active sport participants competing at the World Masters Games, highlighting similarities and differences in visitor experiences among discrete groups of competitors. Visitor experiences are shaped and influenced by the sport and tourism systems at a destination (Figure 8.2).

The spatial distribution and accessibility of sports facilities and venues in central and satellite areas influence the sport experience for both hosts and guests. Bale (1982) makes reference to 'sports nodes' that are functionally delineated areas where the sport experience takes place. These nodal areas can be managed to minimise negative impacts such as congestion, noise and unruly behaviour. Equally, as Smith (2010) notes, 'sport city' zones may be developed to enhance the sport and tourist experience. Similarly, 'Fan Fest zones' (Smith, 2010) which were a feature of the 2006 FIFA World Cup in Germany, and 'experiential hot spots' (Pettersson & Getz, 2009) which may be micro-locations within the local event setting, are defined in space and time.

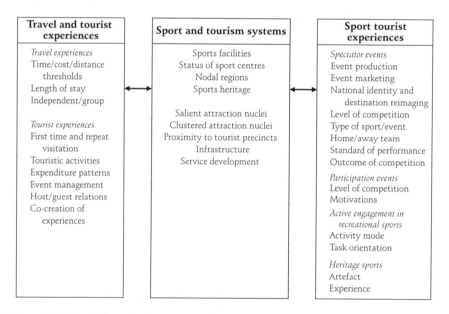

Travel and tourist experiences	Sport and tourism systems	Sport tourist experiences
Travel experiences Time/cost/distance thresholds Length of stay Independent/group *Tourist experiences* First time and repeat visitation Touristic activities Expenditure patterns Event management Host/guest relations Co-creation of experiences	Sports facilities Status of sport centres Nodal regions Sports heritage Salient attraction nuclei Clustered attraction nuclei Proximity to tourist precincts Infrastructure Service development	*Spectator events* Event production Event marketing National identity and destination reimaging Level of competition Type of sport/event Home/away team Standard of performance Outcome of competition *Participation events* Level of competition Motivations *Active engagement in* *recreational sports* Activity mode Task orientation *Heritage sports* Artefact Experience

Figure 8.2 Factors influencing the sport tourism visitor experience

The 'tourism system' that serves sports nodes is equally important. Tourists often embark on a trip with one attraction, or one particular experience in mind (Leiper, 1990). In sport tourism, as in any other form of tourism, tourists typically engage in combinations of attraction nuclei that are salient to the desired experiences of the tourist. The importance of the tourism system in fostering the visitor experience is emphasised by Leiper's (1990) concept of 'clustered nuclei'. Clusters of symbiotic attraction nuclei are a significant element of the contemporary tourism system. Leiper (1990: 375) explains that 'tourists' precinct seems a useful expression for describing a small zone within a town or city where tourists are prone to gather because of clustered nuclei with some unifying theme'.

The development of stadiums within tourist precincts illustrates the potential for tourist experiences to be fostered in association with event sport tourism (Mason *et al.*, 2008; Stevens & Wooton, 1997). A unifying force exists where sports facilities are developed in association with sports bars, museums, halls of fame and other forms of entertainment. The development of complementary activities (e.g. non-sport entertainment) and tourist services (e.g. transport nodes, accommodation, banking and information services) enhances the status of the tourist precinct. These developments may or may not be permanent. The temporary creation of tourist precincts is a strategy that attempts to leverage sports events to encourage a wider experience of the destination. Nash and Johnston (1998), for instance, describe the development of exhibitions, promotions and community events in the cities of Liverpool and Leeds, which hosted games in the 1996 European Football Championship.

Conclusion

This chapter considers the short-term temporal dimension of sport tourism development. It addresses how sport may influence the frequency, timing and duration of sport tourism experiences, and how different aspects of sport and tourism mediate visitor experience at a destination. One of the pressing challenges in the academic study of sport tourism is to develop insights into the relationship between tourist motivations and expectations, and tourist experiences of sport and their behaviours at the destination. This should be informed by an understanding of the complex socio-structural factors that shape and influence tourist experience of sport, and an appreciation of how different factors drive tourist experiences of spectator events, active engagements in participation sports and the heritage values associated with nostalgic experiences. Gibson (1998a–c) originally conceived these as three distinct forms of engagement in sport tourism. The reality is that the interplay of elements of event spectatorship, event participation, active engagement in recreational sports and heritage, in a way that is collectively co-created, offers the potential for powerful and transformative contemporary tourist experiences (Morgan, 2007).

The factors that feature in the recall phase of the sport tourism visitor experience also remain poorly understood. These factors are certain to differ between sport tourist types. High performance athletes are likely to be influenced by the standard of training facilities, personal performance and the outcome of the sport contest (Francis & Murphy, 2005; Maier & Weber, 1993). The experiences of event spectators may be judged by casual spectators based upon the uniqueness of the sport experience while more serious sports fans may judge it by the outcomes of competition and the enhancement of social identity and self-concept (Gibson, 1998a). By contrast, those pursuing general holidays with some incidental engagement in sport may seek unique touristic experiences of sport at the destination (Glyptis, 1982). In each instance, recollections of the visitor experience are influenced by different factors. The sport tourist experience remains a field that is rich in opportunities for further research. Indeed, the rapidly evolving and expanding world of virtual sports and sports experiences that are augmented by virtual reality raises fascinating questions regarding definitions of sport, virtual tourism and new dimensions of the sport experience.

9 Seasonality, Sport and Tourism

It is not known with any certainty whether tourists travel in peak season because they want to, because they have to, or because they have been conditioned to.

Butler, 2001: 19

Introduction

Seasonality is the midpoint on the sport tourism development temporal framework that structures Part 4 of this book (see Chapter 1). In the context of this chapter, seasonality is defined as 'a temporal imbalance in the phenomenon of tourism, which may be expressed in terms of dimensions of such elements as numbers of visitors, expenditure of visitors, traffic on highways and other forms of transportation, employment and admissions to attractions' (Butler, 2001: 5; see also Koenig-Lewis & Bischoff, 2005; Martin *et al.*, 2014). It is one of the most common characteristics of tourism, yet probably one of the least understood. More often than not, it is uncritically viewed as a problem that needs to be fixed. In contrast, seasonality is not generally seen as a major issue in sport, although recent trends towards longer playing seasons tend to constrain multisport participation (Higham & Hinch, 2002a).

The purpose of this chapter is to examine the ways in which sport tourism is characterised by seasonality with an emphasis on how sport has been used to alter tourism seasons in targeted destinations. We start by examining seasonal patterns and issues in a tourism context followed by the sporting context. Consideration will be given to the factors that influence these patterns including the growing implications of climate change on seasonality. Next, a review of sport-based strategies that have been used in an attempt to alter seasonality will be presented. Case Study 9.1 by Robert Steiger highlights the way that climate change is impacting the ski industry and its implications for seasonality in the sport tourism sector.

Seasonal Patterns and Issues in Tourism

Baron's (1975) pioneering study of tourism seasonality consisted of an analysis of tourism data from 16 prominent tourism destination countries covering a period of 17 years. His work confirmed 'most statistical series of arrivals and departures of tourists, bed nights in accommodation, employment in hotels and other branches of the tourist industry show considerable fluctuations from month to month due to seasonality and other predictable factors, which can be measured' (Baron, 1975: 2). Spain has been a particularly popular focus for research on tourism seasonality with Lopez Bonilla *et al.* (2006: 255) identifying a number of different seasonal tourism patterns found in different regions across the country: 'The Andalusia and the Valencian Community display single-peak seasonality; the Balearics and Catalonia have a multiple-peak pattern; and the Canary Islands and the Community of Madrid display non-peak seasonality'. More recently, Martin *et al.* (2014) found that the coastal areas of Andalusia experience the greatest seasonality in Spain, which is further complicated by the fact that they are the most visited areas in the region. Such analyses demonstrate a significant breadth of patterns and the fact that they can differ dramatically by region even when they are in relatively close proximity to each other.

Tourism seasonality tends to be more exaggerated in peripheral areas than urban areas (Jeffrey & Barden, 2001). One of the reasons for this is that the central place characteristics of urban areas mean that there is a greater concentration of year-round attractions in cities (Daniels, 2007; Koenig-Lewis & Bischoff, 2005). These attractions include museums, art galleries, historic buildings, shopping and entertainment venues, many of which are indoor facilities that offer protection from the natural elements. Sporting events, facilities and programmes represent a significant part of this suite of attractions (Chapter 5). In contrast, peripheral areas are characterised by a much narrower range of attractions (Lima & Morais, 2014) that often involve outdoor activities and are therefore more sensitive to weather and climatic conditions. They are also remote by definition, which may present a variety of access problems at certain times of the year (Baum & Hagen, 1999; Cannas, 2012).

Seasonality as a problem

The prevailing view of tourism seasonality is that it is 'a problem to be overcome, or to be "tackled" at a policy, marketing and operational level' (Cannas, 2012: 42). Advocates of this position point out that tourism seasonality has many negative effects on the destination including: (1) economic challenges related to uneven cash flow, employment issues and underutilised capital (Gomez-Martin, 2005; Jang, 2004; Nadel *et al.*, 2004); (2) social issues like crowding during the peak season (Koenig-Lewis & Bischoff, 2005); and (3) ecological issues caused by exceeding carrying capacities during the high season (Chung, 2009; Martin *et al.*, 2014).

Different regions of the world report many of the same problems associated with seasonality, despite experiencing quite different patterns of seasonal variation. For example, Great Britain's high season occurs in the summer months of July and August, with a marked decline in tourism during the winter months. Jamaica, in contrast, has a busy winter season but a slow spring season. Seasonality tends to be more of an issue for destinations that depend on specific climatic conditions as attractions (e.g. winter sports, summer sports and certain adventure sports) than destinations where climatic conditions provide an environmental context rather than the central attraction (e.g. ethnic tourism) (Gomez-Martin, 2005).

With a few notable exceptions (e.g. Chung, 2009; Flognfeldt, 2001), little attention has been paid to the possible benefits that may be attributed to seasonality. Hartman (1986: 3132), however, argues that tourist low seasons offer 'the only chance for a social and ecological environment to recover fully. A dormant period for the host environment is simply a necessity in order to preserve its identity'. Similarly, Butler (1994: 335) suggests that '…while areas may experience very heavy use during peak seasons, in the long run they may well be better off than having that use spread more evenly throughout the year'. In fact, the off-season has been described as having a 'fallow effect' in that it offers the destination a period of recuperation (Baum & Hagen, 1999; see also Koenig-Lewis & Bischoff, 2005).

Seasonal Patterns, Trends and Issues in Sport

Typically, team-based sporting seasons begin with training camps and exhibition games followed by a regular season of league competitions, and ending with playoffs that determine the champion for that particular year. While these patterns vary by sport, level of competition and other factors such as location, one of the most significant changes to sport seasonality over the past 30 years has been the expansion of traditional competition seasons. The reasons for this expansion include an assortment of technological innovations, changing social conditions and general forces of globalisation, and the professionalisation of many sports at the elite level of competition. In conjunction with these factors, partnerships with broadcast media and other corporate partners have generated pressures to increase the length of the competitive season as a strategy for maximising business profits (Sage, 2016). In many cases,

> the restrictions of functioning within a traditional sports season have… been cast aside. The professional development of numerous sports, where teams compete virtually year round has, in those cases, largely eliminated the notion of sport seasonality [in terms of inactivity or an absence of engagement]. (Higham & Hinch, 2002a: 183)

European football is one of many examples that illustrate this trend. The professional football season in Europe has been transformed through the development of international league competitions, from a domestic winter sport into an international club sport that takes place across most of the calendar year. Other examples exist where sport seasons have been altered to revolve around the summer rather than the winter months. The Norwegian Football League, which takes place in summer to exploit favourable playing and spectator conditions, is one example. Similarly, the development of the Super League realigned Rugby League from a winter to a summer sport in the UK and France as part of a strategy to develop a global competition season involving teams based in the northern and southern hemispheres.

Figure 9.1 illustrates the expansion of the rugby union competition season in New Zealand from 1975 to 2017. Using data drawn from New Zealand Rugby Union statistics published on a season-by-season basis in the New Zealand Rugby Almanac, the expansion of the rugby season is evident. This expansion of the sport season was advanced by the introduction of the National Provincial Championship (NPC) in 1976 and the transition to professionalisation of rugby union in the southern hemisphere in 1996 (Higham & Hinch, 2002a). Ongoing league expansion and adjustment to accommodate television broadcast interests continue to extend and shift seasonal patterns (Higham, 2005). Features of the 2010 New Zealand rugby calendar (see also Figure 9.1) included the expansion of the early season Super 12 competition to 14 teams in three countries and ultimately to

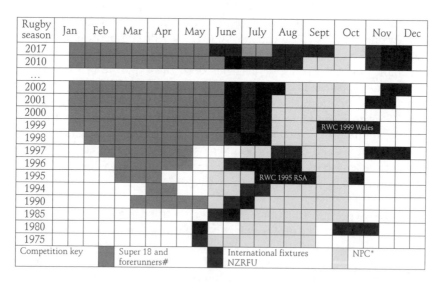

Figure 9.1 The expansion of the New Zealand representative rugby season (1975–2002, 2010, 2017) (Source: New Zealand Rugby Almanac 1975–2002, 2010, 2017)

Note: Years prior to 1994 and from 2002 are not continuous (Source: Higham & Hinch, 2002a).
#Super 18 (2016–2017), Super 15 (2012–2015), Super 14 (2009–2011), Rugby Super 12 (1996–), Super 10/CANZ (1990–1995). Includes preseason warm up games.
*NPC: National Provincial Championship (now overlaps with Rugby Championship (international); RWC: Rugby World Cup.

18 teams in five countries (New Zealand, South Africa, Australia, Argentina and Japan) in 2016. It also included a further expanded international rugby calendar, specifically the expansion of the Tri-Nations competition (from four to six games per team) which extended through August, and the establishment in recent years of the 'Autumn tests' played in Europe (November/December), with a feature test match played in an Asian or North American city en route to Europe in early November each year.

Such has been the continuing expansion of the international programme, with the Tri-Nations evolving into the Rugby Championship with the inclusion of Argentina, that a mid-season 'international window' is squeezed into the latter part of the Super 18, causing that competition to be temporarily suspended for four weeks in June, and recommenced and completed in July. Meanwhile, the northern and southern hemisphere rugby nations continue to seek agreement on a global international rugby season, while the Super 18 will be reduced, once again, to 15 teams in acknowledgement of the unsustainable nature of the current 18-team competition. The relentlessness of the professional rugby calendar in the southern hemisphere has contributed to some leading players seeking occasional respite to prolong their playing careers. This may include the taking up of shorter, more lucrative club contracts in countries such as Japan and France (often in the seasons between the Rugby World Cup tournaments that take place at four-year intervals). More recently, the most high profile players in New Zealand have negotiated contracts that include the option of periods of sabbatical leave of up to eight to nine months to take a complete break from the almost year-round playing season.

The development of all-season sports facilities represents another change that has facilitated the extension of sport seasons. Examples include summer skiing facilities in Scandinavia, the all-season Millennium Stadium in Cardiff (Wales) and the proliferation of smaller-scale leisure sport facilities that effectively provide climatically controlled environments or ones dominated by new technologies (Bale, 1989). For sports that are conducted outdoors, an assortment of equipment and clothing innovations has expanded the range of climatic conditions in which they may be comfortably pursued.

Notwithstanding these changes, it is evident that seasonal patterns still exist in sport. This is most obvious in the case of winter sports such as those that require snow or summer sports such as sailing and scuba diving which are much more attractive to participants in warm water conditions. The reality of these sport seasons has a direct impact on the seasonality of sport tourism.

Sport as a factor of tourism seasonality

At a general level, tourism seasonality has been attributed to 'natural' and 'institutional' factors (Baron, 1975; Cannas, 2012; Koenig-Lewis & Bischoff, 2005; Martin et al., 2014). Natural seasonality refers to regular temporal variations in natural phenomena, particularly those associated with cyclical climatic changes throughout the year (Butler, 1994; Gomez-Martin, 2005). These variations impact demand as well as supply. For example, climate is of fundamental importance to sport tourism in higher latitude destinations, although it is often considered as a nuisance factor or constraint to tourist development. Kreutzwiser (1989) contends that

> Climate and weather conditions ... influence how satisfying particular recreational (sport) outings will be. Air temperature, humidity, precipitation, cloudiness, amount of daylight, visibility, wind, water temperature, and snow and ice cover are among the parameters deemed to be important... In summer, air temperature and humidity can combine to create uncomfortable conditions for vigorous activities, while wind and temperature in winter can create a wind chill hazardous to outdoor recreationists. (Kreutzwiser, 1989: 2930)

These climatic variations are closely correlated with other cyclical events in the natural realm, such as plant growth, animal behaviour and water flows and a range of other destination characteristics that may have direct or indirect impacts on sport tourism activities.

By contrast, institutional factors reflect the social norms and practices of society (Hinch & Hickey, 1996; Koenig-Lewis & Bischoff, 2005). These include religious, cultural, ethnic, social and economic practices as epitomised by religious, school and industrial holidays. Two of the most prevalent institutional constraints on the scheduling of sport travel are school and work commitments (Cannas, 2012; Martin et al., 2014). Tradition also plays a large part in the scheduling of these vacations. Changing religious views, social norms, transportation options and technological advances may moderate these forces.

Butler (1994, 2001) identified three additional causes of seasonality. The first of these is social pressure or fashion, which is usually set by celebrities and other privileged classes within society. A sport example of this factor would be media attention given to celebrities at yachting regattas and horse racing meets. Inertia or tradition is a second seasonality factor (also Cannas, 2012). People tend to be creatures of habit and if they have traditionally taken their holiday during a given time of the year, they will likely continue to do so. For example, even upon retirement, many individuals will take an 'annual vacation' during the same period that they were previously constrained to due to their jobs. Finally, the scheduling of sporting seasons is a factor in its own right. Butler (2001) makes the case that sport seasons have a direct impact on tourism seasons. Winter sports such as skiing, snowboarding and snowmobiling are perhaps the most obvious examples, but summer-based activities such as surfing and golf also influence travel patterns as tourists search for the best seasonal conditions for the pursuit of their sporting passions. Climatic conditions appear to be influential in all of these examples, yet even sports that are played within climatically controlled settings, such as competitive basketball, normally have distinct seasons. If an inclusive definition of 'institutional' determinants of seasonal patterns were adopted, then sporting season would seem to be closely associated with this category. Indeed, sport sociologists have long argued that sport is a social institution (Giulianotti, 2016).

Butler (2001) has suggested that it

is the interaction between the forces determining the natural and institutionalized elements of the seasonality of tourism in both the generating and receiving areas as modified by actions of the public and private sector which creates the pattern of seasonality in tourism that occurs at a specific destination. (Butler, 2001: 8)

Natural and institutional factors can be thought of as pull and push factors which interact with each other to create seasonal patterns (Butler & Mao, 1996; Cannas, 2012; Koenig-Lewis & Bischoff, 2005). While the interactions between the factors that influence seasonal patterns of sport tourism visitation are complex, their basic relationship is relatively straightforward (Figure 9.2). Institutional and natural factors influence tourism demand as well as tourism supply. Policymakers, planners and managers may intervene in this process by modifying supply attributes through strategies such as the development of climatically controlled sport facilities. They may also modify institutional and natural factors on the demand side through such strategies as promotional information that dispels misconceptions that potential tourists might have about sport participation during the off season. It remains unclear as to whether institutional factors can be manipulated at a societal level or whether they will prove to be intransigent. If the latter response is the case, a strategic approach to the manipulation of these factors will be challenging.

Differential influence across the hierarchy of sport attractions

The degree to which a sporting activity influences tourism seasonality is in part determined by the placement of that sport within the traveller's hierarchy of attractions (Chapter 2). Trip behaviour varies on the basis of the centrality of sport as the tourist attraction or how prominently the sport

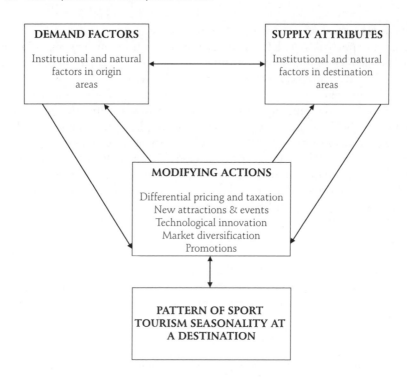

Figure 9.2 Sport influences on patterns of tourist seasonality (Source: Based on Butler, 2001)

features as a travel motivation. Where sport is the principal focus of the trip (i.e. the primary attraction), travellers demonstrate a greater propensity to travel in the tourism 'off-season' (World Tourism Organisation and International Olympic Committee, 2001). More casual sport tourists (i.e. those who see sport as a secondary or tertiary attraction) exhibit higher levels of seasonal variation in their travels. Where sport is the primary motivation, sport tourists are willing to negotiate institutional and natural constraints that might otherwise be insurmountable for more casual sport tourists (Hinch & Jackson, 2000; Hinch et al., 2001). For example, outbound tourists from Germany, the Netherlands and France who have a strong sport focus are more likely to travel from January to April than are outbound tourists with a more casual approach to sport. The latter group is much more likely to travel from May to August (World Tourism Organisation and International Olympic Committee, 2001). In contrast, casual sport tourists exhibit a single-peaked pattern of seasonality that coincides with the typical summer peaks of mass tourism in these countries. A similar pattern was found for sport tourists in Canada (Weighill, 2002).

The degree to which a sporting activity is dependent on specific climatic conditions is also an important factor in terms of seasonal patterns of activity (Gomez-Martin, 2005). Sports such as skiing and sailing are directly tied to specific natural attributes such as snow and wind conditions, respectively. Other sports, even though they occur outdoors, may be enhanced by natural attributes but these attributes are not necessarily central to the experience. In the latter case, natural factors like weather may serve as the general context rather than having a direct bearing on the essence of the tourist experience (Scott, Jones & Konopek, 2007). In cases such as mountain biking and beach volleyball, natural conditions which deviate from what is perceived as the ideal may be a significant deterrent to sport tourism at certain times of the year. Alternatively, advantageous or extreme weather conditions (e.g. unseasonably warm temperatures leading to snow melt and high river flow volume) may be promoted as a positive characteristic for many types of extreme sports (e.g. white-water rafting, kayaking).

Climate Change and Sport Tourism Seasonality

Climate 'is the prevailing condition of the atmosphere deduced from periods of observation' while weather is 'the state of the atmosphere in a given place at a given time' (Gomez-Martin, 2005: 572). Tourists experience climate as weather at any point in time and space. Until fairly recently, the tourism industry tended to view the natural causes of tourism seasonality as stable but climate change is increasingly recognised as a dynamic force in the temporal and spatial distribution of tourism (Hartman, 1986; Kennedy & Deegan, 2001; Koenig-Lewis & Bischoff, 2005). The trend towards rising global temperatures has impacted tourism seasonality in a variety of ways. Amelung et al. (2007) used two climate change scenarios with the Tourism Climate Index (TCI) to show that climatic conditions currently preferred by tourists will shift to the higher latitudes. Visitation to locations like the Mediterranean will likely shift from the summer peak season to the spring and fall seasons while Scandinavia (and other) destinations in the higher latitudes will be characterised by a longer summer peak season. Similarly, Fang and Yin (2015) used the TCI to demonstrate that China currently enjoys climatic conditions that support a summer peak in the north, a winter peak in the south and a bimodal shoulder peak (spring and fall) in the middle latitudes. They warn, however, that the impacts of climate change on these seasonality regions are likely to be profound and that more research is needed in this area.

A study of the potential impacts of climatic change on winter recreation in Ontario, Canada, highlights the vulnerability of four major winter sport activities (Scott et al., 2002). Even relatively small increases in temperature were shown to result in major decreases in Nordic skiing, snowmobiling, ice fishing and downhill skiing activities. The least affected of these activities was downhill skiing due to the availability of snow-making equipment that increases the range of temperatures for which snow cover can be guaranteed. Notwithstanding this technology, a relatively small increase in temperature was shown to reduce the average ski season by between 21% and 34%. Given the dispersed nature of the other activities, current snow-making technology was not seen as a feasible way to mitigate the potentially drastic impacts of global warming in what is currently a winter-based destination.

The broad-based shift of preferred tourist climates to the higher latitudes with its implications for seasonality is only part of the story. Specific sport tourism activities like downhill skiing depend on particular aspects of climate such as temperature and precipitation (Tuppen, 2000). Scott, McBoyle & Minogue (2007) suggest that the desired conditions for such activities will not only shift to the higher latitudes but will shift to the higher alpine elevations (Case Study 9.1). Despite a variety of management strategies designed to address these issues, the ski season in most locations is shifting.

The relative degree to which the seasonal impacts of climate change are manifest in a spatial (e.g. shifts to the higher latitudes) or in a temporal (shifts across the calendar year) manner is difficult to predict and will be regionally variable. To a large extent, this depends on whether the institutional factors that shape tourism seasonality discussed earlier are flexible or not. If these factors prove to be inflexible then climate change is likely to have a spatial impact on tourist flows, whereas if institutional factors are flexible, a temporal impact on travel flows is more likely to occur (Amelung et al., 2007).

Case Study 9.1: Climate Challenges and Responses by the Ski Industry

Robert Steiger, University of Innsbruck

Snow is an indispensable resource for mountain destinations in the winter season. Climate change alters the amount of snowfall (decrease) and snow melt (increase), influencing the average snow depth and the length of the snow season. Any impact of climate change on this basic resource is a potential stressor for snow-dependent resorts. But the impact differs between latitude and elevation and also within relatively small regions due to the climatic heterogeneity of mountain regions (Steiger, 2010). Winter seasons with an extraordinary lack of snow in the late 1980s resulted in average demand losses (participation drops) of 30% in ski areas in Switzerland and northern Italy (Abegg & Frösch, 1994; Steiger, 2011a). At that time, snow-making was not widespread and therefore the ski areas were highly dependent on natural snow. Due to the critical challenges presented by snow-scarce seasons, ski areas started to invest more and more in snow-making. Consequently, demand losses due to snow-scarce winters in the 2000s were considerably lower than in the previous two decades, e.g. −10% to −12% in Canada, the United States and Austria (Dawson *et al.*, 2009; Rutty *et al.*, 2017; Scott, 2006a, b; Steiger, 2011b).

Nowadays, snow-making is the most used measure to adapt to climate variability and climatic changes. But, several other adaptations have been used in winter destinations over time. The Olympic Winter Games is an exemplary case. Rutty *et al.* (2014) identified three climatic adaptation eras: emergent adaptation (1924–1956), technological transition (1960–1984) and advanced adaptation (since 1988). At the beginning of the emergent adaptation era, all sporting events were held outdoors and were dependent on natural snow and ice. Hockey and figure skating were the first events to be moved indoors during that era. Other weather risk adaptations were large-scale snow transfers from snow stockpile sites to the event locations, often supported by the army, and systematic use of climate data and weather insurance. The technological transition era is characterised by the increased use of mechanised weather risk management technologies. These include mechanical snow compaction, refrigerated indoor ice surfaces for speed skating and concrete bobsled and luge tracks with built-in refrigeration systems. Indoor competition venues for ice hockey and figure skating became standard during this era. The 1980s Winter Olympic Games in Lake Placid (USA) were the first to use snow-making for alpine and Nordic events. In the advanced adaptation era, major technological adaptations developed in previous Games were standardised and expanded. Snow-making was used not only to cover 100% of alpine and Nordic slopes, but also to stockpile snow several months before the Games. For the Sochi Games 2014, snow was produced and stockpiled in the previous winter season and even stored over summer. Refrigeration systems are also increasingly utilised, e.g. to cool the ski jump track (since 2010). More events moved indoors (speed skating, curling) and the use of weather services has intensified through an increasing number of installed weather stations providing site-specific hourly weather data and forecasts.

The main reason for all these adaptations is to reduce the risk of cancellations which was becoming more important as visitor numbers and broadcast revenues increased from US$0.5 million in 1960 to US$1.2 billion in 2010 (Rutty *et al.*, 2014). But, the technological adaptations also enabled warmer cities to host the Games. During the era of technological transition, average temperatures of the host city in February were 3.1°C higher than in the emergent adaptation era. In the advanced adaptation era, average February temperatures in the host city were 7.8°C higher than in the first era, with Sochi being the warmest

(Continued)

Case Study 9.1: (Continued)

city that ever hosted the Olympic Winter Games in 2014, considering both the city and the locations of the events (Rutty *et al.*, 2014).

Consequently, important questions are if and where these adaptations can cope with climate change. Scott *et al.* (2014) investigated the ability of the 19 previous host regions to provide suitable conditions in the future. The number of reliable regions decreases to 11 (low emission scenario) and 10 (high emission scenario) in the 2050s, with recent host locations and unsuccessful bidders of the recent Games being at high risk (Oslo, Vancouver) or unreliable (Garmisch/Munich, Sochi). In the 2080s low emission scenario, 10 of the former locations would still be reliable, whereas in the high emission scenario only six reliable destinations would remain viable. These results suggest that more weight should be given to the climatic suitability of potential future locations to host the Winter Olympic Games and that climate change needs to be taken into account, especially as almost a decade lies between the bid process and the actual hosting of the event, during which time further climatic changes will inevitably accelerate and advance.

Adaptations to weather and climate risks carried out for mega-events like the Winter Olympic Games represent the technical possibilities of the snow-based winter tourism industry. Some of these measures have already been adapted for use in mass tourism winter destinations to buffer and ameliorate changes in the seasonal availability of snow. Snow-making has become standard practice of ski areas throughout the world. Stockpiling snow and storing it over the summer season is an emergent strategy to secure an early season start for Nordic and alpine skiing. Kitzbühel in Austria for example, a prominent potential victim of climate change due to its low elevation, uses several snow depots to be able to start the ski season between mid and late October. In the 2016/2017 winter season, it had the longest non-glacier ski season in Austria. This trend of early ski openings has led to a paradoxical situation: the seasonal evolution of demand has decoupled from seasonal snow availability. On the northern hemisphere, the demand peaks are during the Christmas–New Year holidays and semester/winter break (typically in February). But the snow depth maximum on the mountain is reached between March and May, depending on elevation and latitude. It is evident that further technical adaptations to secure the snow product require greater financial capacity. Climate change is thus likely to intensify competition among ski areas with likely market contraction, resulting in potential winners and losers in the fight to remain seasonally viable in increasingly challenging climatic circumstances.

Climate change also potentially increases seasonality, as from an operational perspective in order to prevent losses, the same turnover must be generated in a shortening ski season, leading to higher demand peaks. This is even more problematic for destinations that are almost entirely dependent on the winter months. Therefore, winter destinations have started to vitalise the summer season by introducing new products (e.g. mountain biking, family fun parks on the mountain, staged theme hiking trails) to attract customers in the summer season. But, positive effects on the economic viability of businesses in these resorts and on social aspects (e.g. year-round jobs instead of seasonal jobs) are accompanied with negative environmental effects, as more tourists in a sensitive mountain environment increase disturbances in the ecosystem.

Key Reading

Scott, D., Steiger, R., Rutty, M. and Johnson, P. (2014) The future of the Olympic Winter Games in an era of climate change. *Current Issues in Tourism* 18 (10), 913–930.

Steiger, R. (2011b) The impact of snow scarcity on ski tourism. An analysis of the record warm season 2006/07 in Tyrol (Austria). *Tourism Review* 66 (3), 4–15.

Additional references cited in this case study are included in the reference list.

Strategic Responses

Tourism managers, planners and policymakers have addressed seasonality issues in numerous ways. Responses have included attempts to lengthen the main season and/or establish additional seasons by diversifying markets, providing tax incentives to suppliers at targeted times of the year, scheduling staggered school holidays, encouraging distinct domestic and international seasons, introducing differential pricing at different times of the year, adjusting supply inventories and introducing new festivals and conferences during periods that traditionally experience low visitation (Baum & Hagen, 1999; Butler, 2001; Parrilla *et al.*, 2007; Koenig-Lewis & Bischoff, 2005; Lee *et al.*, 2008). These strategies have been hampered by the relatively low level of understanding of the complex interactions that contribute to seasonality and the lack of a sound theoretical understanding (Amelung *et al.*, 2007; Koenig-Lewis & Bischoff, 2005). This limitation can be addressed in part by adopting Weaver and Lawton's (2002) typology of strategic approaches to addressing tourism seasonality along with Hinch and Jackson's (2000) application of leisure constraints theory to tourism seasonality.

A strategic typology

Weaver and Lawton (2002) identified six basic strategic responses to seasonality that have been adopted by several other researchers (Jang,

2004; Koenig-Lewis & Bischoff, 2006). These strategies are divided into three demand-based approaches (increase, reduce or redistribute) and three supply-based approaches (increase, reduce and redistribute). In practice, demand and supply approaches are closely related (Cannas, 2012) and are combined in the following discussion.

The first set of strategies focuses on increasing visitation outside the peak season. This strategic approach favours urban over peripheral areas as it seeks to establish a critical mass of attractions across all seasons (Koenig-Lewis & Bischoff, 2005). Product/market development and diversification lie at the heart of this approach with the hosting of sporting events and festivals being one of the most popular strategies (Getz, 2008). A good example of product/market diversification through the introduction of sporting events is provided by the Isle of Man (Baum & Hagen, 1999). The Isle of Man has traditionally been a popular summer tourist destination in the UK, but a sharp decline in the 'sun seeker' markets in the early 1980s prompted the development of a product diversification strategy designed to increase sport tourism in the shoulder seasons. Existing sporting events such as the Manx TT road race scheduled during the shoulder tourism period of late May and early June served as the anchors for the introduction of other related events. Assessments of the Manx TT show that it attracts approximately 37,000 visitors per year who spend in excess of £15 million at the destination. This major event is supported by a series of other motor sport events throughout the year that each attract between 3000 and 6000 visitors. Another example of a sport-based strategy designed to increase visitation outside the peak period is the provision of additional indoor sport facilities such as spas, swimming pools and climbing walls.

Conversely, in response to the prospects of shorter ski seasons, downhill ski resorts have developed golf courses and mountain biking trails as a way of attracting more visitors during the summer months (Hudson & Hudson, 2016) (Focus Point 9.1). In the face of falling visitation in the 1980s, ski resorts in North America and Europe made a concerted effort to improve and diversify their product (Tuppen, 2000). One of the key improvements was a major expansion in snow-making equipment which allowed heavily used runs between the upper slopes and the base of the resort to open sooner and close later in the year, thereby facilitating an extended season. Just as importantly, it built consumer confidence within active sport tourism markets that there would be snow at the resorts during what had previously been considered a very marginal period. Many resorts also expanded the range of winter-based activities that visitors can partake in through the provision of indoor sports and fitness centres, and facilities for other types of winter sports such as snowboarding, cross-country skiing and snowshoeing. Large resorts like those found in Whistler, British Columbia, Canada, developed summer attractions like golf courses to become all-season destinations (Hudson & Hudson, 2016).

Focus Point 9.1: All-Season Ski Resorts in British Columbia

All-season resorts have become a common strategy to combat seasonality in the ski industry. British Columbia, Canada, is home to some of the best examples of these types of strategies. One of these is Panorama resort located two hours to the west of Banff. Strategies to increase visitation outside of the ski season have included the spring, summer and fall scheduling of events like the '1000 Peaks Triathlon, the Mad Trapper BC Cup Mountain Bike race, the SEA2Summit adventure race, and the Shimano Dirt Series Women's Skill Camp' (Hudson & Cross, 2005: 198). To the north – just outside of Jasper National Park – the proposed Valemount Glacier Destination is also being touted as an all-season ski resort but one with a difference. At Valemount, the resort will be located so that it provides access to year-round skiing on a glacier (Oberti Resort Design, 2016). But the best example of the transformation of a ski resort into an all-season resort in BC is Whistler Blackcomb in the coastal mountains just north of Vancouver. In addition to being one of the world's premium ski and snowboarding destinations, the area has become a popular destination during the rest of the year. While resort operators invested in snow-making technology to extend the ski seasons into the late fall and early spring, they not only faced technical limitations associated with warm daily temperatures but they also had to deal with institutional constraints inclusive of societal traditions that see Canadians shift to summer activities when the temperatures in their home areas are warm. Notwithstanding good ski conditions in the early to late spring at Whistler, the Vancouver market was interested in summer activities. Adopting the adage that if you 'can't beat them then join them', Whistler Blackcomb began hosting regular festivals throughout the summer, developing world-class golf courses and building extensive cycling networks that took advantage of the beautiful mountain landscapes. The downhill cycle trails even use existing ski lift infrastructure and have made a major contribution to the area, becoming a mecca for warm weather adventure sports (Whistler Blackcomb, 2017).

Key Reading

Hudson, S. and Cross, P. (2005) Winter sports tourism destinations: Dealing with seasonality. In J. Higham (ed.) *Sport Tourism Destinations* (pp. 233–247). London: Elsevier Butterworth Heinemann.

Additional references cited in this case study are included in the reference list.

Golf developments have also been used effectively to modify seasonal visitation in other tourism contexts. For example, Baum and Hagen (1999: 309) reported that Prince Edward Island, Canada, successfully developed golf tourism as a way to encourage visitation during the spring and fall. This was done through a planned strategy that featured public and private sector investment and targeted promotions to senior and retired markets interested and able to visit during the shoulder seasons. On a more cautionary note, Garau-Vadell and Borja-Sole (2008) described the success of a similar strategy in Mallorca, Spain. However, their longitudinal study found that the growth rate for golf has slowed, the use of tour packages inclusive of hotel bookings has declined and marketing channels have shifted from traditional sources to internet-based practices. Their findings highlight the fact that individual sports are dynamic and may fall in and out of popularity (see Chapter 10). This dynamic suggests that careful assessment of a sport facility diversification strategy is required, especially in the case of peripheral destinations, to verify that an adequate return on investment will be achieved. The lower costs of product diversification through a new events strategy explain the popularity of this approach over more capital-intensive redevelopment strategies.

The demand side of these strategies includes accessing new but complementary market segments (Cannas, 2012; Koenig-Lewis & Bischoff, 2005). In a tourism seasonality context, there are a number of market segments that are traditionally recognised as having fewer constraints relating to the timing of travel. These groups include senior citizens, conference delegates, incentive travellers, empty nesters, affinity groups and special interest tourists (Baum & Hagen, 1999). Another form of market diversification addresses the institutional constraints that sport markets face at different times of the year. A good example of this type of approach to the resolution of a seasonal visitation problem is illustrated by Eurocamp's use of geographic market segmentation (Klemm & Rawel, 2001). Eurocamp specialises in self-drive holidays in Europe that feature active sporting amenities at campgrounds en route. Initially, the company targeted British families but due to the institutional constraint of school holidays, bookings were concentrated in August. A conscious strategy to promote their product to other European countries that had different school holiday periods was successfully pursued over a 15-year period. The outcome of the market diversification strategy was consistent, with stable visitor flows from May to September rather than the single high season month of August. Jang (2004) proposed a variation of a market diversification strategy through the application of financial portfolio theory. He argued that marketers should select a mix of tourism segments that fall along a Seasonal Demand Efficient Frontier given a demand–risk target. At the very least, destination marketers should identify the strategic priorities in terms of these segments.

The second set of strategies for addressing the challenges of tourism seasonality aim to *reduce visitation during the peak season* and can be as simple as increasing prices or restricting access to publicly owned attractions (Cannas, 2012; Koenig-Lewis & Bischoff, 2005). To a large extent, the market will respond to excess demand in the short term with an increase in prices. In destinations where local carrying capacities are being exceeded during the peak season, it is important that development controls are in place to limit the entry of new suppliers. This has been accomplished in Canadian Rocky Mountain National Parks by strictly controlling the capacity of accommodation stock and services that are available to tourists (Ritchie, 1999). From a sport perspective, an active policy to not host major sporting events during peak visitation periods and to limit access to sport infrastructure such as secured recreational areas during the peak season will contribute to this strategy. The modification of institutional factors such as distributing school holidays more evenly throughout the year can also be pursued (Cannas, 2012). Similarly, traditional sport seasons can be adjusted. While many of these types of adjustments must come at a state, national or even international level that is beyond the direct influence of destination managers, other strategies can be pursued at a local level. For example, the local destination can position itself as a training site for selected sports outside of the normal tourist high season (Yamaguchi *et al.*, 2015).

Finally, the third set of strategies to address the challenges of tourism seasonality is to *redistribute existing visitation* more evenly throughout the year. By rescheduling sport events and by capitalising on the unique seasonal patterns of individual sports, it is possible to manipulate seasonal distributions of tourism visitation in a destination. These approaches can be combined with differential pricing and target marketing such as promotions designed to shift visitors from the high season to other seasons (Koenig-Lewis & Bischoff, 2005). In publicly managed recreational areas, a temporal variation of the popular spatial zoning practices can be implemented. Under this approach, access restrictions can shift as required on a temporal as well as a spatial basis.

Leisure constraints theory

Notwithstanding the successes reported above, seasonal variations in tourism remain a prominent feature of the industry. The 'stubbornness' of these patterns suggests that a stronger theoretical understanding of seasonality is needed (Amelung *et al.*, 2007). Leisure constraints theory represents one framework that provides additional insight into this area (Hinch & Jackson, 2000; Hinch *et al.*, 2001). This theory considers the barriers that prevent non-participants from taking part in leisure pursuits as frequently as they would like. In the context of sport tourism seasonality, leisure constraints theory raises the question, 'What is it that inhibits people from travelling for sport at certain times of the year?'. The answers to this question provide a better understanding of sport tourist seasonal behaviour and would identify constraints that can be targeted by managers.

Hudson *et al.* (2010) used a leisure constraints framework to examine the barriers to snowboarding and skiing in Canada. Their framework is positioned in a tourism context and operationalises the hierarchical model of leisure constraints (Crawford *et al.*, 1991; Walker & Virden, 2005) to identify management options designed to increase participation in downhill skiing. For example, one of the constraints that they found was that non-skiers were afraid that they would be cold and uncomfortable on the slopes. A logical management response to this is to raise consumer awareness of technological advances in the manufacture of winter clothing that will allow enhanced ski comfort. Figure 9.3 has been modified from the hierarchical model of leisure constraints (Jackson *et al.*, 1993) to emphasise its relevance in the study of sport tourism seasonality.

One of the key characteristics of this model is the order in which seasonal constraints are encountered and negotiated. In the context of sport tourism seasonality, sporting preferences are the starting point. A major consideration at this stage is the centrality of seasonal factors in terms of the motivations for travel. In cases where natural seasonal factors like climatic conditions are the primary attraction or motivation for the sport tourists (e.g. sunny warm conditions for casual sport tourists, favourable snow conditions for serious snowboarders), their absence in a destination during a given time of the year will be a major and perhaps insurmountable constraint. Jackson *et al.* (1993) have labelled these as intrapersonal

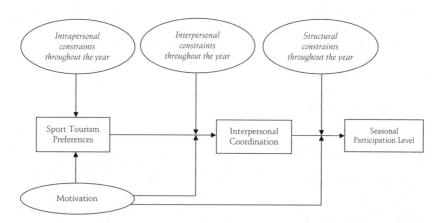

Figure 9.3 Hierarchical model of seasonal sport tourism constraints (Source: Based on Jackson *et al.*, 1993)

constraints. Where these constraints do not exist, sport tourists proceed to the interpersonal constraint level. Team sports, or those that are enhanced by fellow participants, require the potential sport tourist to coordinate his or her travel plans with others. Potential sport tourists who seek travel and sport companions but cannot find them will likely not participate even though they have the initial motivation. Finally, structural constraints consist of things such as high travel costs, lack of accommodation and school or work commitments. Ways around this last level of constraints may be negotiated by the potential sport tourist, although they too may prove to be insurmountable. Innovative packaging by the host destination can facilitate this negotiation process throughout the year. By understanding this leisure constraints framework, sport tourism destinations can identify appropriate seasonal target markets and help them to negotiate the particular constraints that they face at various times of the year.

Conclusion

Both tourism and sport are characterised by seasonal variations during the course of a year. Whereas these variations are generally seen as undesirable in a tourism context, there has been relatively little concern about seasonal variation in a sporting context. Somewhat paradoxically, seasonal patterns of tourism tend to be stable despite ongoing attempts to modify them, while sporting seasons have undergone considerable change – especially in terms of their extension throughout the year. A variety of natural and institutional factors influence seasonality in both of these realms, and it is increasingly recognised that climate change is having an impact on tourism seasonality. This chapter addressed the role of sport in seasonality and its potential to address the challenges of climate change and tourism seasonality more strategically.

It is clear that sport is not only a factor in tourism seasonality but that it can potentially be harnessed as a means to modify tourist visitation patterns over the course of a year. Lengthening competition seasons for spectator sports and scheduling sport events during tourism shoulder seasons are direct ways of increasing visitation to a destination during these periods. Active sport tourism also represents an opportunity to consciously influence tourism seasonality in a destination. This is especially true for sport tourists who are highly motivated members of sport subculture groups. Destinations should consider the resources that they have in their region during the tourism shoulder seasons that may be attractive to these groups. Destinations that offer a unique blend of attributes can adopt a niche marketing strategy that effectively attracts a geographically dispersed but passionate group of visitors during non-peak times of the year. Finally, while nostalgia sport tourists have not been discussed in depth in the chapter, the tourism literature on seasonality highlights the potential for museums and similar types of facilities to attract visitors outside of the summer months (Stevens, 2001).

While tourism managers have long pursued the economic benefits of modifying seasonality, sport managers are increasingly adopting a similar approach. There are financial benefits for sports that attract visiting participants at non-traditional times of the year. While some of these benefits may be collected directly at the gate of the sporting venue, others can be leveraged through the tourism industry. For example, sport tourists travelling during the off-season are likely to enjoy lower rates for accommodation due to higher vacancy rates. These benefits can be maximised if collective action on the part of a particular sport can demonstrate that there are a substantial number of sport-related visitors arriving during this period. Group rates and adjoining travel packages can then be negotiated.

The positive side of tourism seasonality mentioned at the beginning of this chapter should not be lost in the search to solve the 'problem' of seasonality. Sport tourism destinations can benefit from a 'fallow period' that allows for the regeneration of natural and human resources. Notwithstanding the tendency for many sport seasons to expand schedules to the point where they are almost year-round pursuits, there are downsides to this. Spectators and athletes may burn out if they do not have a chance to re-energise during an off-season. Sports offer their most powerful function as tourist attractions when the enthusiasm of participants is at its peak – yet this peak cannot be prolonged or sustained indefinitely. In the interest of the sustainability of the sport, sport tourism and the destinations where this activity takes place, it may be strategically prudent to maintain some form of seasonal variation over the course of a year.

10 Evolutionary Trends in Sport Tourism

In some respects, it was tourists [to France] who thus passed on mountaineering to sports enthusiasts in the 18th and 19th centuries, before the latter, the mountaineers, then offered tourist skiing in return during the 20th century.

Bourdeau *et al.*, 2002: 23

Introduction

This chapter examines the dynamics of sport tourism over the long term. It opens with a discussion of tourism and sport life cycles. Adam Doering's case study of the evolution of surfing and its implications on travel patterns highlights the interrelationship of these life cycles (Case Study 10.1). Next, we consider heritage-based sport travel, which represents a special type of interaction in which sport heritage and nostalgia serve as a tourist attraction. Examples include travel to visit sport halls of fame, sites of past sporting events and imagined pasts played out through fantasy sport camps and programmes. Neither the interaction between life cycles nor heritage sport tourism occurs independently of other forces operating in the broader environment. The chapter therefore closes with a focus on the web of global trends that are likely to influence the future of sport tourism.

Intersecting Life Cycles in Sport and Tourism

Destination and product life cycles are dominant features of tourism (Butler, 2006; Chapman & Light, 2016). Butler's (1980) tourist area life cycle model epitomises this idea with six stages: exploration, involvement, development, consolidation, stagnation and either rejuvenation or decline. In revisiting Butler's model, Johnston (2001a) suggested that the early part of this cycle could be classified as the pre-tourism era in that some other institutional framework besides tourism dominates the destination while the latter stages of stagnation and decline can be described as a post-tourism era. In between these points, a series of mini tourism product cycles can occur in the destination. The general pattern of these cycles is one of increasing visitation until the destination's resources are adversely affected or the attraction simply falls out of favour, at which point visitor numbers begin to decline (Chapter 4). A variety of implications emerge from these cycles with the most important being that management intervention is needed to sustain tourism resources if the destination's life span is to be extended.

Sport attractions play a significant role in the life cycles of many destinations. For example, in his detailed analysis of the destination life cycle of Kona, Hawai'i, Johnson (2001b) noted that the original 'Ironman' race was transferred from Honolulu to Kona in the early 1980s. This shift corresponded to the last period of Kona's development phase and served as an image-maker for the destination. It marked a critical point in the development of Kona by replacing the 'way of life' destination image with an 'active sports' image. Similarly, surfing in the Miyazaki Prefecture played a major role in revitalising a local tourism industry in this part of Japan (Doering, 2017). Successfully positioned as a 'southern tropical paradise' from the 1930s through to the 1950s, Miyazaki transitioned into a honeymoon resort in the 1960s and 1970s. Surfing was introduced in the 1960s and slowly gained momentum throughout the next four decades. By the end of the 1970s, honeymoon tourism had shifted to other destinations. Over the next three decades, the surf culture in Miyazaki evolved organically and by 2010 there was a demonstrable convergence of surf culture and tourism that positioned Miyazaki as the pre-eminent surf destination in Japan. With the inclusion of surfing as a demonstration sport in the 2020 Tokyo Olympics, Miyazaki's status as a surf destination will continue to develop and evolve over time (Case Study 10.1).

Case Study 10.1: From *He'e nalu* to Olympic Sport – A Century of Surfing Evolution

Adam Doering, Wakayama University

On 3 August 2016, the International Olympic Committee's (IOC) unanimously voted to introduce surfing as a demonstration sport in the Tokyo 2020 Games. For the first time in history, 20 female and 20 male surfers will compete for medals in Chiba, Japan, as part of the Olympic competition. The inclusion of surfing in the Games was anticipated over 100 years earlier as Olympic swimming gold medallist and father of modern surfing, Hawaiian Duke Kahanamoku, noted in his biography, 'Even as early as…(1918), I was already thinking of surfing in terms of how it could someday become one of the events in the Olympic Games. Why not?' (Kahanamoku & Brennan, 1968: 37). A century later, the more pertinent questions may be: Why now? Why Japan? What political, social and economic factors enabled this dream to become a reality at this particular point in history? To address these questions, this case study traces the evolution of surfing and its interrelationship with tourism over time before reflecting on the possible future that surfing's inclusion in the Olympic Games invites.

Surfing has its roots in pre-colonial Hawaiian and Pacific Island history (Ingersoll, 2016). In the Hawaiian language, surfing is referred to as *he'e nalu*, meaning 'wave sliding' (Ingersoll, 2016). With the earliest recorded history dating back to the mid-1700s, *he'e nalu* was a widespread social, political and spiritual activity, replete with surf chants, boardmaking rituals and other sacred elements embedded within everyday practice (Ingersoll, 2016). *He'e nalu* was, and still is, a form of ocean literacy, connecting Hawaiians to the rhythms of the tides and currents as well as helping to navigate the seafaring histories that comprise the Hawaiian Islands and its people (Ingersoll, 2016). As an integral part of the pre-colonial Hawaiian every day, *he'e nalu* was a way of knowing and being in the world defined by a dynamic relationship with the sea.

By the end of the 19th century, Western colonisation of the Hawaiian Islands introduced a combination of disease and Christian morality that would almost eliminate *he'e nalu*. The precarious (re) birth of modern surfing in the early 20th century was directly related to the colonisation of Hawai'i when, in 1898, American business interests annexed the islands. In an effort to attract investment and boost tourism in Hawai'i, *he'e nalu* was commodified and repackaged as an idealised, tropical island fantasy fit for visitor consumption. *He'e nalu* turned into surfing and had become a tool for tourism development. The Waikiki Beachboys, including Duke Kahanamoku, played a critical role in the early emergence of surf tourism development, not only teaching tourists to surf in the slow rolling Waikiki waves, but also reviving the traditions of *he'e nalu* in the more heavy breaks of the North Shore of Oahu (Ingersoll, 2016). The tradition of *he'e nalu* was preserved, but a new mode of surf tourism had also emerged.

Surfing eventually departed Hawai'i, arriving on the California coast in 1907 and Australia in 1915 (Warshaw, 2010). In 1907, a surfing demonstration was held by Hawaiian-born George Freeth in Redondo Beach, California (Moore, 2011). The surf demonstration was designed to promote a new beach culture and encourage real-estate development along the California coast. A core group of surfers soon picked up the sport and began experimenting with new board designs and materials. By 1950, polystyrene began being used to make the longboards lighter to transport and easier to ride, making surfing more accessible. The release of the Hollywood blockbuster *Gidget* (1959) introduced surfing's subculture to the general public, inspiring a whole new generation to head down to the beach (Warshaw, 2010). It is estimated that prior to the release of the film there were roughly 5000 surfers, but by 1963 estimates of the number of surfers ranged between 150,000 (Finney & Houston, 1996) and 2 or 3 million (Moore, 2011). Surfing had become mainstream.

(Continued)

Case Study 10.1: (Continued)

Surf films continued to play an important role in reshaping the sport as it expanded. As the popularity of surfing grew, beaches became overcrowded and another surf ideal began to surface: the search for the perfect wave. *The Endless Summer* (1964), a film about two quirky surfers travelling the globe in search of the world's best waves, embedded the idea of surf travel into the mainstream discourse. Travelling surfers began to identify as 'free surfers' or 'soul surfers' as they sought out waves to ride in remote areas of Indonesia and the Pacific. Finding the perfect wave and surfing it expanded the sport into new locations and by the late 1970s a global commercialised surf tourism industry had been established.

Decade by decade the sport grew and slowly transformed. In the 1970s, Australia's surf culture started to leave its mark, playing an important role in professionalising, commercialising and transforming surfing into a competitive sport. Competitive surfing flourished alongside the sport's global incorporation, which led to the development of some of surfing's biggest Australian brands – Quiksilver, Billabong and Rip Curl (Stranger, 2010). In turn, competitive surfing inspired technological innovation as competitors sought to gain advantage through improved board design. The shortboard revolution in the late 1960s and the manufacturing of the three-fin 'thruster' boards in the 1980s allowed for sharper turns, greater acceleration and a more aggressive surf style to emerge. These key innovations enabled surfing to become the high-performance competitive sport known today.

Surf corporations began promoting major competitions all over the world and sponsored the sport's top competitors. Simultaneously, the same surf corporations began sponsoring other highly talented 'free surfers' living the alternative travelling lifestyle. The free surfer marketing, with professionals travelling to exotic locations to surf in sponsored videos and feature in corporate promotions, encouraged the further commercialisation of the surf travel lifestyle established in the 1960s (Stranger, 2010). With the arrival of globalisation, surf culture, surf tourism and the corresponding surf economy have now breached the shores of even the most isolated waves. Transnational surf corporations are worth billions of dollars. Today's surfing has become fully incorporated into the consumer capitalism regime.

Over the past century, surfing has multiplied and mutated into a broad range of meanings and practices. Surfers are riding waves in German rivers, constructing wave pools in Texas and sliding across stormy swells of Lake Superior. Surfers are riding longboards, tandem, skimboards, hydrofoil boards and standup paddleboards. There are lifestyle surfers, recreational surfers, competition surfers, soul surfers, drifters, organic yoga surfers, big wave surfers, surfers who avoid crowds and surfers who feed off the energy of a crowd. More women are surfing than ever before and the waves are filling up with teenagers training to improve their surf performance levels. Importantly, local Hawaiian surfers continue to engage with *he'e nalu* as part of the ongoing struggle for autonomy within a neocolonial surf tourism destination (Ingersoll, 216). All of this is surfing in the new millennium.

Despite this diversity, the dominating framing of surfing's future centres on the sport's inclusion in the Olympics. As with its modern genesis in Hawai'i a century ago, the incorporation of surfing into the Olympics could be understood as another attempt to capitalise on surf culture in an effort to develop broader commercial interests in the emerging markets of Asia and beyond. Japanese authorities have stressed that the Olympic competition will be held in the sea to promote marine sport tourism. Japanese surfers likewise hope that the professionalisation of surfing will provide an opportunity to gain respect from mainstream Japanese society, but are also concerned that the increased exposure will overwhelm the already overcrowded breaks (Doering, 2018). The inclusion of surfing in the Tokyo 2020 Olympics consequently offers a glimpse into the near future of the sport: global expansionism, surf tourism destination development, overcrowding, high performance sport progression, and the advancement of surfing as a spectator sport. The far future of surfing, however, will no doubt emerge from the growing tension between its increasing diversity and ambitious plans to turn surfing into a universal, lucrative sporting phenomenon.

Key Reading

Doering, A. (2018) Mobilising stoke: A genealogy of surf tourism development in Miyazaki, Japan. *Tourism Planning & Development* 15 (1), 68–81.

Stranger, M. (2010) Surface and substructure: Beneath surfing's commodified surface. *Sport in Society* 13 (7–8), 1117–1134.

Additional references cited in this case study are included in the reference list.

The evolutionary dynamics of sport

If sports are to be used as a mechanism to rejuvenate tourism destinations, it is important to understand how they typically evolve.

Like tourism products, individual sports, sports disciplines and sport events have their own life cycles. They too go 'out of fashion'. And they increasingly find themselves having to compete against other leisure activities and events... In sport too there is a constant need for the adaptation of individual sports and events to the changing requirements of sportsmen and sportswomen, as well as spectators. (Keller, 2001: 4, 5)

Just as a primary measure of the tourism destination life cycle is the number of visitors to a destination, a primary indicator of sport life cycles is the number of participants and spectators involved. Other measures of the status of a sport within its life cycle include the sophistication of rule structures, the level of skill development and physical performance and, increasingly, the extent of commodification and professionalisation.

The dynamics of sport have been clearly illustrated in recent years by the slower growth of many highly structured competitive team sports and the ascent of individualised and extreme sports (Breivik, 2010; Gilchrist & Wheaton, 2016; Klostermann & Nagel, 2014; Wheaton, 2004). This evolution began in earnest at the beginning of this millennium with Keller (2001) pointing out that at that time:

> The membership for organized types of sport is on the decline, as are the proving grounds from which top-level sports traditionally draw new blood. The new generation is a sliding, gliding and rolling generation. Their sports are freestyle events like inline skating, street basketball and snowboarding, which in many cases are associated with a youthful subculture. Performance and rankings no longer play any role. What counts are the aesthetic, 'feel-good', atmospheric effects. (Keller, 2001: 13, 14)

Breivik (2010) characterises these emerging sports in terms of risk, participation sites inclusive of demanding natural and constructed environments, loose organisation, distance from the dominant sport culture and individual participation within developing subcultures. Such sport is also manifest in the growing differentiation found in traditional sports like mountain climbing (e.g. indoor climbing, bouldering, sport climbing) (Focus Point 10.1); the emergence of air sports (paragliding, hang gliding and sky diving through to base jumping, sky boarding and acrobatics); new board sports (snow-, skate-, wakeboarding); and variations of bicycling (trick, mountain, BMX), skating (in-line) and luge (street luge). It is safe to assume that innovative forms of sport will continue to emerge although the individual popularity and longevity of each of these sports is more difficult to predict.

Focus Point 10.1: Climbing – Growing Differentiation

Cory Kulczycki, University of Regina

Rock climbing began as one of the elements of the sport of mountaineering (Nettlefold & Stratford, 1999). Since its birth as an independent activity, it has continued to grow and evolve with diversification and specialisation being the norm rather than the exception. Variations in rules and practices now distinguish several different types of rock climbing including but not limited to traditional climbing, sport climbing, aid climbing, top roping and bouldering (Levey, 2010). In traditional climbing, the climber places anchors (nuts and cams) into cracks in the rock, which connect to a rope thereby providing protection from falls; this protection is removed when the climb is completed (Steele, 2006). Sport climbing emphasises speedier gymnastic-style movements and uses bolts that are permanently placed within the rock (Schuster *et al.*, 2001). The aid climber is assisted by various tools while climbing (Abramson & Fletcher, 2007). Top-roping uses anchors at the top of the climb through which the rope passes to the climber and partner (i.e. the belayer). Bouldering involves climbing short routes/problems on a cliff or large boulder without the use of ropes, harnesses or other climbing equipment (Levey, 2010). These styles of climbing differ based on how climbers approach the activity, their behaviours and their perceptions of authenticity (Kiewa, 2002). Climbing styles, sites, social interactions, past experiences and cultures all influence the travel decisions of outdoor climbers with sites like Squamish, British Columbia, having a particularly strong reputation for bouldering (Kulczycki, 2014).

Recently, there has been an upsurge in the number of indoor climbing facilities designed to mimic or deviate in appearance and function from traditional outdoor spaces. These facilities are proving to be popular for training, socialisation and fitness in urban areas (Eden & Barratt, 2010). The increasing popularity of these facilities fosters different experiences and skill sets and has led to passionate debates among the climbing community about authenticity and place meaning (Eden & Barratt, 2010; Kulczycki & Hinch, 2015). One of the most important questions in a sport tourism context is whether the built facilities will become popular travel destinations just as their outdoor counterparts have proven to be?

Key Reading

Kulczycki, C. (2014) Place meanings and rock climbing in outdoor settings. *Journal of Outdoor Recreation and Tourism* 7–8, 8–15.

Kulczycki, C. and Hinch, T. (2015) 'It's a place to climb': Place meanings of indoor rock climbing facilities. *Leisure/Loisir* 38 (3–4), 271–293.

Additional references cited in this focus point are included in the reference list.

The reasons for the shift to individualised and more hedonistic sport are manifold. In their study of changes in German sport participation, Klostermann and Nagel (2014) suggest that societal shifts in values such as a move towards individualism and a growing interest in health and body culture have led to the growth of individual sports. Breivik (2010) articulated three other possibilities with the first being 'compensation'. As some societies become more controlled and safety-oriented, some individuals – especially youth – will look to extreme sport as an antidote or a countermeasure to boredom. Alternatively, this shift may be seen as an extension or 'adaptation' of a modern life that features more sensory stimulation as manifest in areas such as entertainment, cuisine and travel. Seeking variation in sport may be a simple extension of the variation sought in these other areas of life. Finally, a third possible explanation is that the variation in modern and post-modern society may be seen as superficial and often virtual (e.g. computer gaming). In response, extreme sport provides participants with the opportunity to express themselves in a real way though their bodies – not just their minds.

One of the dilemmas facing extreme sport participants is the tendency for such sports to evolve from counterculture to mainstream activities. Typically, these sports have emerged for people who wish to get away from 'someone else's' rules and regulations, and set up or be part of their own autonomous renegade groups. Yet, these sports are themselves part of evolutionary processes. As sports institutions, media, equipment and clothing manufacturers and the tourism industry interact, extreme sports tend to shift from subculture to mainstream (Breivik, 2010; Hoffer, 1995). This evolution introduces structures and rules that serve to ensure that the activity is managed in a way that facilitates commodification (Wheaton, 2013) until some group once again breaks away to begin something new. Although they recognise this pattern, Gilchrist and Wheaton (2016: 25) point out that young lifestyle sport participants are both consumers and producers of sports. As such, they have considerable ability to 'fashion their own cultures, identities and experiences in ways that are never fully determined by adults, public authorities, corporate interests, or socio-cultural norms'.

Snowboarding provides a good illustration of this process. It emerged as a subculture activity in resistance to the dominant culture of alpine skiing which had become mainstream by the 1980s. It was characterised by a non-traditional view of sport. Yet, the initial radical nature of snowboarding has been steadily moderated through the pressures of commodification. The development of snowboarding as a commercial television product by ESPN illustrates this point. Another benchmark in the evolution of snowboarding from a subculture to a mainstream sport was the inclusion of snowboarding in the 1998 Winter Olympic Games. Other emerging issues involve the relationship of these activities to nature. While initially seen as eco-friendly, increasing extreme sport participation in remote areas is causing environmental stress. At the same time that remote natural areas are under pressure, new urban-based extreme sports such as parkour (the physical practice of traversing urban elements) are becoming popular. More generally, extreme sports tend to be dominated by Western male youth and it will be interesting to see whether this changes significantly in the future. It should also be noted that an increasing number of extreme sports are being invented and/or sponsored by corporate entities like Red Bull. While the participants in many of these sports can be considered a sporting subculture, they are subcultures that build the brand and markets for these aggressive corporate entities (Gorse *et al.*, 2010).

The evolutionary dynamics of tourism

At a global level, tourism is relatively unique in that it has demonstrated consistent growth with the interplay of domestic and international travel. Butler (2006) has characterised this growth in terms of cycles (e.g. life), waves (e.g. successive cycles) and wheels (cycles within waves) at a destination level. While he cautions that chaos theory suggests that prediction is difficult, he also points out that tourism has continued to grow despite major disruptions like global economic downturns, terrorist attacks and natural disasters. This resilience appears to be rooted in the high value that people place on leisure in their personal lives and in the diversity of destinations and activities that populate the tourism system.

In the context of this consistent growth, tourism cycles and waves have a significant influence on the development of sport. For example, over the past two decades, golf has been introduced to many warm climate destinations as a tourism development strategy (Garau-Vadell & de Borja-Sole, 2008; Humphreys, 2014). In the process of this development, opportunities to participate in the sport of golf have been extended to and taken up by local residents. A reciprocal relationship also exists in that tourism provides

> an opportunity for leisure activities to be popularised. With increased popularity they have developed into formally organized sporting activities. Some even progressed from leisure activities to Olympic disciplines. Beach volleyball and snowboarding are two good examples of this. (de Villers, 2001: 13)

Tourism not only introduces sport to new locations, it also fosters innovation within sport. Figure 10.1 illustrates that change in sport occurs in recreational and competitive settings, the former being more conducive to major innovation than the latter. Both settings are influenced by external trends associated with the economy, environment, politics, society and technology. Sport innovations often originate outside of the home environments. Recreational sport tends to be more conducive

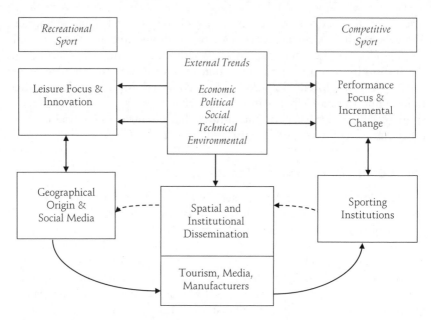

Figure 10.1 Innovation in sport (Adapted from Keller, 2001)

to innovation than institutionalised sport because experimentation is encouraged in most leisure and tourism settings. Keller (2001) convincingly argues that the change in location and uninterrupted free time that tourists enjoy while on holiday is conducive to innovation in sport pursuits.

In competitive settings, the focus is on performance in terms of recognised physical skills and strategies. Rules and practices are structured in ways that discourage radical change and encourage conformity. Major innovations in recreational sport settings are shaped by leisure patterns that may be unique to specific geographic regions and which are characterised by spontaneity and freedom. In contrast to the relatively unconstrained settings in a recreational context, a variety of formal sporting institutions actively inhibit variation and change in competitive sport environments. The spatial and institutional dissemination of major innovations and incremental change in sports occurs through social media, the traditional media and by the marketing efforts of sporting goods manufacturers. Tourists also act as significant agents of innovation and dissemination. They introduce new sporting interests to tourist destinations and may, in turn, bring home insights into new sports that they were exposed to while travelling.

Heritage Sport Tourism

Like sport tourism, heritage tourism is recognised as a major category of tourism activity as well as an important realm of academic endeavour (Timothy & Boyd, 2006). Ramshaw and Gammon (2007, 2016) suggest that at the nexus of these realms is the phenomenon of heritage sport tourism. Other authors such as Gibson (1998a–c) have used the intriguing although somewhat narrower concept of nostalgia-based sport tourism.

Nostalgia sport tourism positions sport heritage as a tourist attraction (Focus Point 10.2). It provides tangible evidence of the way that sport life cycles can have a direct impact on tourism in the form of sports halls of fame and museums, high profile sporting venues and a range of thematic programmes, all of which take on the mantra of tourist attractions in their own right (Delpy, 1998). Fundamentally, nostalgia '…is a bittersweet emotion that involves the desire to go back to the past…' (Cho *et al.*, 2014). As such, sport has always been closely associated with nostalgia, perhaps because it reconnects people to their youth when they were typically more active (Gammon, 2002). Snyder (1991) suggests that sport nostalgia can trigger reflections on our own mortality, which often results in an idealisation of the past.

> On the surface it appears that halls and museums attract people because of their fascination with sport, including the idolized figures and memorabilia from the past. But this is only part of the explanation; the attraction may also be based on the contrasts in incongruity between past and present. This juxtaposition of the past with the present creates the context for feelings of nostalgia. (Snyder, 1991: 229)

Past sporting experiences may become reference points from which sport-oriented people derive life meaning. This meaning results from both collective and individualised views of the past. In the case of the former, popular media and various sporting institutions celebrate an assortment of sporting victories, events and personalities in a way that impresses them upon the popular consciousness. In the latter case, sport nostalgia is linked to the benchmarks of an individual's sport involvement and identity at different points in his or her life. It often is reflected in their social media practices. Fairley and Gammon (2005) point out that this nostalgia is not just focused on sport memorabilia or other tangible manifestations of sport heritage but is, in part, due to the desire to relive a social experience that one may have had related to sport. The combination of collective and individualised nostalgia creates a powerful force that is increasingly being mined by the sport and tourism industries in order to create commercial development opportunities.

Focus Point 10.2: Heritage, Nostalgia and Outdoor Ice Hockey

Gregory Ramshaw, Clemson, University

Falla (2000: 54) argues that outdoor ice hockey represents a 'facet of northern recreational heritage'. Outdoor rinks, in particular, take on many forms, such as the backyard rink, the rural, farmyard pond and the urban, community rink. In many parts of North America, outdoor rinks have been ascribed mythical and symbolic qualities. As Ramshaw and Hinch (2006) argue, rinks are the places that take us back to our childhood, where unencumbered play was more important than rigid, codified sport; where expressions of nationalism – particularly in Canada – could be performed; and where the winter was not something to be feared, but embraced. Similarly, the outdoor rink became a reactionary nostalgic symbol to the highly competitive state of amateur ice hockey, as it became an emblem of an egalitarian past where age, ability and economic circumstances were seemingly erased – although, as Ramshaw (2010a, 2010b) argues, the historic reality of outdoor rinks was not particularly equal, specifically in terms of gender. Given strong ties to nostalgic and heritage pasts ascribed to outdoor rinks, they are now reproduced in many ways: in marketing campaigns and commercials, in artwork and memorabilia, in large spectator events and in participation events. Most notably, outdoor ice hockey events have become staples of the National Hockey League (NHL) annual schedule. These events are typically played on artificial ice hockey rinks in large outdoor baseball and football stadiums inclusive of props and staging such as old-fashioned jerseys and snow-covered winter landscapes. Between 2003 and 2018, 25 NHL outdoor hockey games were played at stadiums across Canada and the United States, with the attendance ranging from around 30,000 (Winnipeg in 2016) to over 100,000 (Ann Arbor, Michigan, in 2014). Moreover, during the past 15 years, outdoor ice hockey spectator events have also been embraced by minor league hockey leagues, college hockey leagues and European hockey leagues. Well over 100 non-NHL outdoor games have been played in venues as far reaching as Red Square in Moscow to the Pula Arena, a 2000 year-old Roman amphitheatre, in Croatia. As such, the frequency and proliferation of spectator-based outdoor hockey events have likely reduced both the media and touristic appeal beyond local regions, as these are no longer 'once in a lifetime' and 'must see' spectacles (Ramshaw, 2014b). On the other hand, the significant growth of regional and national 'pond hockey' tournaments such as the World Pond Hockey Championships (Lowes & Awde, 2015), which are played on natural frozen bodies of water such as lakes and ponds and feature a far-less codified and structured form of the sport commonly referred to as 'shinny', speaks to another sport tourism dimension. This form of outdoor ice hockey can be categorised as 'participation-based events' at which personal and collective heritages and nostalgia are kinaesthetically embodied and performed by sport tourists.

Key Reading

Ramshaw, G. (2014b) Too much nostalgia? A decennial reflection on the heritage classic ice hockey event. *Event Management* 18 (4), 473–478.
   Additional references cited in this case study are included in the reference list.

As could be expected, people of middle age or older are often seen as the primary cohorts for nostalgia-based tourism. Snyder (1991) suggested that

> ...for many people sport triggers feelings of longing for the past when they had pleasant experiences associated with sport. This reflection is most evident for the middle-aged and elderly, which have had more sport experiences, but perhaps more important, this is a period of their lives when concern about their own mortality is salient in their self-reflections. Consequently, for those involved in sport, nostalgia may provide a source of consolation and a means of adjustment to the uncertainties of their lives. (Snyder, 1991: 238)

Gammon (2002) counters this point by arguing that nostalgia is also of interest to youth as part of a popular culture that draws on the past as a way of establishing 'new' trends. Sport museums and halls of fame are the primary manifestation of nostalgia sport tourism (Ramshaw, 2010a, 2010b). The National Baseball Hall of Fame located in Cooperstown, New York, is a good example. It was opened in 1939 and attracts approximately 400,000 visitors per year (Gammon, 2002) with a running total that has surpassed 14 million visitors. The induction of new members is a high profile annual media event in the United States.

Past, current and, in some cases, future sites (e.g. designated Olympic sites) of sporting events and activities are a second major type of nostalgia sport tourism attraction (Ramshaw & Gammon, 2010). These sites have an inherent appeal as special places where heroes played and legends were made (Stevens, 2001). Such an aura fosters an emotional nostalgic experience, which focuses on the connection between place and sporting performance. However, tensions between heritage and modernity are common. While modern facilities that enhance the performance of athletes, the spectator experience and revenue opportunities are sought through the development of new facilities, the loss of sporting place identity is recognised as a high price to pay. Elaborate strategies designed to inject the soul of the old facility into its replacement include orchestrated ceremonies involving former athletes who starred in the old facility, figuratively or literally passing the torch to the current generation of athletes in the new facility (Belanger, 2000). Another approach to capturing nostalgia is the inclusion of sport museums and halls of fame in new facilities. For example, the sport museum at Camp Nou (FC Barcelona – football stadium), which was opened to the public in 1984, hosted its 30 millionth visitor in 2016 (FC Barcelona, 2016).

Fantasy sport programmes are a third and particularly intriguing variation of nostalgia sport tourism (Gammon & Ramshaw, 2007). They range from mock training camps to themed cruises, restaurants and bars. Gammon (2002) suggests that there are five main motivations to travel to fantasy sport camps: (1) the desire to be associated with a famous event; (2) the opportunity to train in a famous or meaningful facility; (3) to increase identification with a particular team or club; (4) to be closely associated with sporting heroes; and (5) general interest in the sport and in skill development. For the nostalgia sport tourist, these camps enable participants to escape from the routine of their day-to-day existence. They provide nostalgia sport tourists with the opportunity to relive experiences from their youth or to reconstruct them by living their sporting dreams rather than revisiting past realities. While these selected manifestations of nostalgia-based sport tourism are prominent, a wide range of variations exists. Cho *et al.* (2014) present a conceptual model of nostalgia in the context of sport tourism to provide further insights into these variations. Their model highlights four types of nostalgia: nostalgia as past personal experience which may be induced by *sport objects* such as stadiums; nostalgia as socialisation in which an individual constructs identity through nostalgia-based *social interactions*; sport nostalgia that is consciously used to build self-identity; and sport nostalgia used to establish group membership and a sense of belonging.

Heritage sport tourism adopts a more critical perspective of sport heritage by recognising that sport has featured negative practices that, while insightful, are not worthy of celebration in the same sense that the nostalgia dimension of sport has been. Beyond nostalgia, scholarship on sport heritage tourism has been used to provide insight into memorialising conflict, marginalising people and illustrating the dissonance between sport heritage and contemporary tourism promotion (Ramshaw, 2014a, 2014b). More broadly, Ramshaw and Gammon (2016) have identified *heritage of sport* as distinct from *heritage as sport*. In the former, the sport attraction draws on the celebration of achievements and events within the sport itself while in the latter, sport is recognised as a reflection of the collective identity of the people who reside in a destination. Both of these perspectives reflect a long-term temporal dimension in sport tourism that goes beyond the limitations of nostalgia.

Major Trends Affecting Sport Tourism

The discretionary nature of sport and tourism makes predictions about the future difficult as they must be made within a broad and complex environment. Trend analysis does, however, provide insights into the likely directions that sport tourism will move toward. Continued growth is projected in the realm of sport tourism events (Getz & Page, 2016) and sport tourism more broadly. However, this growth is unlikely to take the form of a simple linear extension of existing sport tourist participation patterns. While tourism will face an assortment of major challenges and opportunities, the complexity of the tourism system and the environments in which it operates ensure that tourism development will be anything but linear (Buckley *et al.*, 2015) For example, while environmental challenges have been identified as a major constraint to future tourism development, the trend towards an experience-based economy and activity-oriented travel may mitigate the negative impact of the former (Tolkach *et al.*, 2016).

As discussed earlier in this chapter, sport participation has been characterised by a shift from competitive teams sports to individualised and adventure sports (Bourdeau *et al.*, 2002; Breivik, 2010; Gilchrist & Wheaton, 2016; Klostermann & Nagel, 2014; Wheaton, 2004). There are increasing variations within sport (e.g. surfing: short boarding, long boarding, paddle boarding, body surfing) and increasingly sophisticated sport careers that include changing travel patterns as one progresses along a sporting career path (Getz & McConnell, 2014). The reasons for these shifts are to a large extent either an extension of broader societal trends (increased mobility, post-modern production and consumption practices) or a counter-reaction to them (e.g. embodied experiences and exposure to risk) (Breivik, 2010). New sporting opportunities will continue to emerge in urban settings based on demand and technology, but peripheral areas will also grow in stature as sport places of uncertainty and risk. These trends are themselves rooted in the broader context in which sport and tourism exist.

Economic trends

Rising income is the greatest driver of tourism demand and conversely falling incomes and decreasing income security in times of economic uncertainty are powerful determinants of changing tourism flows (Dwyer *et al.*, 2008). Economic surges and downturns will be reflected in sport tourism activity. Positive economic factors for tourism include deregulation, increased trade, improved information technology and dynamic private sectors, while negative factors include cyclical economic downturns, protectionist trading practices and large disparities in growth and development between countries and regions.

Globalisation is perhaps the most dominant economic trend that has emerged from the latter part of the 20th century (Chapter 4; Higham & Hinch, 2009). It has exerted increased pressure for the commodification of sport and tourism. Of particular significance is the tendency towards the convergence of tourism, leisure, sport and entertainment. This is especially true of elite organised sport where the trend towards professionalism and 'show business' is evident (de Villers, 2001; Keller, 2001). Another impact of globalisation has been the emergence of the experience economy, which has resulted in a search for active experiences in a tourism context (Tolkach *et al.*, 2016). The growth of active sport tourism reflects these trends, as does the development of sport events that include active engagement, celebration and communitas (Chalip, 2006).

While there have been several recent popular movements designed to push back against globalisation such as the vote in the UK to exit the European Union and the election of a US president who espouses protectionist trade and immigration policies, it is unlikely that the processes of globalisation will be reversed. In fact, the growth of the Chinese and Indian economies through globalisation processes has made them influential players on the demand side of tourism (Buckley *et al.*, 2015; Tolkach *et al.*, 2016). From a sport tourism perspective, this means that the cultural understandings and motivations for sport that characterise the residents of these countries will have a substantial impact on the distribution and activities of sport tourists. Globalisation has also played a major role in the environmental crises that are outlined shortly, but such impacts also have economic consequences. Buckley *et al.* (2015) point out that one such consequence will be increased oil prices which will translate into fewer long-haul tourist trips as these flights become increasingly costly and less socially acceptable (also see Higham *et al.*, 2016).

The media plays an important role in this process. Popular media has had a long association with sport inclusive of the golden age of newspapers, radio and television. It is the last of these forms of media, however, that is perceived as having the greatest impact on elite sport (McKay & Kirk, 1992). From the outset, the television broadcasting of sport generated concern that fans would stop going to the actual competition in favour of viewing sports from the comfort of their homes. At the root of this concern was the belief that sport revenues would shift from traditional sporting institutions to the broadcasters.

The economics of sport were founded on the principle of persuading large numbers of people to leave their homes, to travel to enclosed sporting venues and to pay for entry in order to view professional performers engage in various forms of structured, physical competition. (Rowe, 1996: 569)

History has shown that broadcasters have indeed enjoyed substantial financial rewards but they have also generated significant financial benefits for owners and administrators of televised sports, as well as the destinations where these sports take place. Television revenues for professional sport now far exceed gate revenues. Despite this changing economic context, there are still concerns that the media has subverted sport for its own purposes and by doing so, it has eroded the integrity of these sports. For example, Rowe (1996) noted that

> Television has progressively exerted pressure on sports to be played at times convenient to broadcast schedules and to modify rules in order to guarantee results, to prevent events going too far 'over time', and to overcome any dull passages that might tempt viewers to reach for the dial (later the remote control). The global spread of sports television has created its own severe pressures on sport by, for example, demanding that wherever possible 'live' sports should be transmitted at a time convenient for the largest and most lucrative TV markets. (Rowe, 1996: 573)

Interactive technology in the form of pay-per-view television, online streaming, social media and video games presents substantial challenges for sport tourism. Over 25 years ago, Johnson (1991) speculated that the increasingly interactive experience of watching sport from the comfort of home may eventually result in the need to pay spectators to attend televised games in order to create an exciting atmosphere in the sporting venue. While this prediction has not yet become a common reality, the distribution of complementary tickets to increase crowd size and atmosphere at events with poor advanced ticket sales is a common practice – especially when a large broadcast audience is anticipated. For sport tourists, the benefits of the on-site experience must clearly exceed the costs associated with the trip. It is also important for sport tourism managers to advocate the retention of the things that make sport distinctive. The inherent authenticity of sport (Hinch & Higham, 2005) provides a competitive advantage to many tourism destinations that may be lost if the nature of the activity shifts to staged entertainment.

The influential role of the media is not limited to mainstream sports. Subculture sports can also be closely tied to the media. In today's post-modern society '...the specialized press ... plays a fundamental role in initiating participants to techniques, equipment, cultural codes and languages which lay the foundations for the identity of the sports "tribes"' (Bourdeau et al., 2002: 27). This media is, therefore, of particular interest to sport tourism managers as it influences where sport subcultures travel in pursuit of their sporting passions.

Environmental trends

Over the long term, the interaction of sport and tourism will be strongly influenced by environmental trends (see Chapters 7 and 9). There are many environmental trends of relevance but climate change is especially important (Buckley *et al.*, 2015). Although there remains much political debate about its measurement, causes and implications, there is widespread recognition of its existence and concern for its implications for tourism and sport (Hall & Higham, 2005). The focus of most of the research in this area is on visitor flows and the distribution of impacts, although more recently, consideration has been given to the role of tourism as a causal factor in the process of this change.

Studies of the impact of climate change on the level and distribution of tourist arrivals and expenditures are generally based on modelling. Projections are made about changing climatic conditions and then translated into maps of shifting distributions of tourist arrivals as tourists seek out their preferred climatic conditions. It has been noted that there will be both winners and losers as a result of this redistribution given that in some locations conditions will improve while in others they will be less attractive (Hein *et al.*, 2009; Iordache & Cebuc, 2009). While tourism in general is subject to these shifts, tourism related to winter- and summer-based sporting activities is seen as being particularly sensitive to these changes (Dodds & Graci, 2009). This is certainly the case in terms of the ski industry with its dependence on snow (see Case Study 9.1). Moen and Fredman (2007) predict a decrease in the number and expenditures of skiers to existing ski resorts in Sweden as snow accumulations decrease. Scott and McBoyle (2007) recognise similar challenges facing the ski industry elsewhere but note that these impacts can be moderated by strategic responses in terms of demand and supply. On the demand side, the response to changing climate conditions can be moderated by improved weather forecasting and reporting along with an assortment of constraint negotiation strategies that skiers can implement (e.g. altering the timing of their ski vacations). On the supply side, a suite of technological practices can also be used to moderate the impact of changing natural snow conditions. These include: (1) the use of snow-making equipment, (2) improved ski slope landscaping and operational practices and (3) cloud seeding. Business strategies include the development of (1) ski conglomerates, (2) revenue diversification and (3) modified marketing practices. A more radical but already existing alternative is the development of indoor ski areas.

Tourism is increasingly being recognised as a significant contributing factor to climate change with approximately 4%–6% of all of the world's carbon dioxide emissions estimated to accrue from tourism activities (Dodds & Graci, 2009). Approximately 70%–75% of this total is attributed to the transport sector with the balance coming from the accommodation, food and beverage and the attraction sectors (Iordache & Cebuc, 2009). The tourism industry is therefore in a position to reduce carbon dioxide emissions proactively or be targeted by regulatory agencies. Unfortunately, the tourism industry has been slow to appreciate the long-term implications of climate change and a deteriorating environment (Tolkach *et al.*, 2016). However, in their exploratory study of the 2010 FIFA World Cup, Otto and Heath (2009) found that hosting a mega-event such as the World Cup increased the level of awareness of tourism and sport stakeholders on the impact of their activities on climate change. This consciousness led to concrete strategies and initiatives designed to reduce CO2 emissions during the event and in subsequent activities.

Political trends

Political stability is a precondition of tourism development (Dwyer *et al.*, 2008; Tolkach *et al.*, 2016). The United States and Europe are likely to remain very influential in global politics but face growing competition from China, India and Russia as these economies grow. Non-state actors (e.g. ethnic, cultural, indigenous and religious perspectives) will become increasingly engaged, perhaps reshaping spatial and temporal elements of sport tourism. Increasing income disparities within and between developing and developed countries will continue to impact the flow and distribution of sport-based tourism. Changing balances of power and influence drive political trends. The continuation of international free trade agendas, despite the protectionist position recently taken by the United States, highlights changing power configurations throughout the world. However, the economic benefits of free trade are unlikely to be evenly distributed. Hall (2000b) predicts that there will be

> increased conflict between developing and developed countries over global economic development strategies as it becomes apparent to large numbers of the population in developing countries that they will never be able to have western lifestyles due to population and resource constraints. (Hall, 2000b: 88)

Similar disparities have emerged within countries as some regions benefit while others suffer as a result of shifting economies under globalisation. The repercussions of such sentiments and other political grievances are already being felt in terms of protest activities that have direct impacts on tourism and sport. Despite these issues of globalisation, and perhaps because of them, there may be positive impacts on tourism and sport as they are used as instruments of geopolitical strategy. Buckley *et al.* (2015) offer the example of the Commonwealth Games sport competition which has been branded as the 'friendly games' being used to unite a disparate membership.

Terrorism is not a new challenge in tourism (Hall, 2000b; Sonmez *et al*., 1999) nor is it alien to sport. Perhaps the most notable terrorism incident in a sport tourism context was the fatal attack on Israeli athletes at the Munich (1972) Olympic Village (Wedemeyer, 1999). Yet, despite the high profile of sporting events as terrorist targets, there have been surprisingly few politically motivated attacks on sport. Indeed, it has been suggested that sport tourism was one of the more resilient types of tourism in the aftermath of the terrorist attacks in America on September 11, 2001 (World Tourism Organisation, 2001). The most obvious implication of terrorist threats at major sporting events is increased costs of security requirements. To illustrate, in 2016 it was estimated that the security costs for the 2020 Tokyo Olympics would be US$1.44 billion which is more than 11% of the estimated US$12.6 billion for the total Games budget. Given the enormity of this budget item and given national interests, the Japanese government will help the local Olympic organising committee to cover these costs (*The Japan News*, 2017). Not only will such high costs make hosting some of these events prohibitive for many cities, it may also lead to an increase in preference for off-site spectatorship over on-site consumption if spectators feel at risk or become frustrated by the security measures. Active sports characterised by dispersed spatial patterns of participation may increase in popularity relative to major spectator or participant-based event sport tourism. Similarly, nostalgic tourism may increase as traditional patterns of event sport tourism consumption are altered.

Social and demographic trends

Socio-demographic trends are also likely to exert considerable influence over the future of sport tourism (Delpy, 2001). Dwyer *et al*. (2008: viii) suggest that 'Tourists are increasingly interested in discovering, experiencing, participating in, learning about and more intimately being included in the everyday life of the destinations that they visit'. Sport offers a mechanism for meeting this desire. In terms of specific demographic trends, immigration has resulted in changing sport activity patterns. This has presented a challenge to sport tourism activities such as the ski industry in Canada where Asian immigrants are much less likely to patronise ski resorts than residents of Anglo Saxon descent (Hudson *et al*., 2010). Similar challenges exist in terms of aging populations in North America.

People born between 1946 and 1964 make up almost a third of the North American population and they started switching to more gentle winter sports such as snowshoeing and cross-country skiing, because their aging bodies could no longer handle the rigours of alpine runs. High-tech computer designed skis and equipment, which make it safer and easier for even aging outdoor lovers to learn or continue to enjoy the sport, have halted that trend somewhat in recent years. (Loverseed, 2000: 53)

In the face of this trend, sport tourism operators have to adjust their products to match the needs of their markets. This adjustment involves a shift from hard adventure activities to less physically demanding soft adventure outdoor sport activities. An aging population will likely become more health conscious and, therefore, seek sport activities that will help them to retain and perhaps regain their physical and mental health rather than put them at risk.

The shift from a modern to a post-modern society is also mediating the context of sport tourism. At one level, this shift has its economic roots in the neoliberal rejection of the welfare state and regulated markets in favour of competition, free trade and globalisation (Dwyer *et al.*, 2008; Stewart & Smith, 2000). Niche markets, individualism, flexibility, time fragmentation, new technologies, innovative communication networks and commercialisation all characterise today's society. The role of place in post-modern sport is also changing. Local tribal loyalties based on the 'home team' are shifting to attachments to corporate identities or brands. Table 10.1 summarises the changing face of organised sport from the modern to the post-modern era.

Table 10.1 A comparison of modern and postmodern sport

Dimension/component	Modern sport	Post-modern sport
Game structure	Rules are sacred	Rule modification and experimentation
Team leadership	Conservative	Adventurous
Values and customs	Amateurism, respect for authority, character building	Professionalism, innovation
Organisation and management	Central control	Diffusion of authority
Financial structure	Gate receipts	Sponsorship, television rights, gate receipts, sport as business
Venues and facilities	Basic seating at stadiums	Customised seating, video support
Promotion	Limited	Extensive
Viewing	Live match attendance	TV/live streaming audiences dominate
Spectator preference	Display of traditional craft	Eclectic blend of entertainment
Fan loyalties	Singular and parochial loyalty	Multiple loyalties – all spatial scales
The sports market	Undifferentiated mass market	Fragmented and niche markets
Coaching and training	Rigid, repetitive practices	Blend of science and naturalistic practices – variety

Source: Adapted from Stewart and Smith (2000).

Technological/communication trends

Technological innovations have irrevocably changed the face of sport and tourism. They have improved sporting performances and enriched tourism experiences. Moreover, technology has further blurred the line between sport and tourism. For example, advances in social media have resulted in the evolution of sport-related web sites from static information repositories to interactive branded sites that often form the basis of a leisure experience in their own right (Gilchrist & Wheaton, 2016). These emerging technologies are breaking down the barriers of time and space. Dwyer *et al.* (2008) identify advances in information and communication technology as having the most important implications for tourism. Fundamentally, they argue that advances in this realm 'provide businesses with the tools to respond to individual preferences and to stimulate tourism purchases' (Dwyer *et al.*, 2008: x). A good example of the power of this new technology is the growing role of social networking in the decision-making of travellers (Xiang & Gretzel, 2010). Similarly, lifestyle sport participants 'make sense of their sport through the sharing of moves, manoeuvres and styles and engage in vibrant discussions on web forums about evolving practise and the deeper philosophical meanings' of their respective sports (Gilchrist & Wheaton, 2016: 24). Such practices have accelerated the dispersion of emerging trends across the globe.

At a more fundamental level, virtual reality and cyberspace technology is having direct impacts on the way that people experience leisure (Buckley *et al.*, 2015). The extent that sport experiences in cyberspace can substitute for sport experiences in real space remains a matter of speculation. Current examples of sport experiences that take place in cyberspace include live online sports commentaries, fantasy sport leagues, online gambling, real-time viewer surveys during sport broadcasts and instant progress updates and live digital video images of sports contests. On another level, many computer games closely conform to definitions of sport (Chapter 2) inclusive of a physical activity component (Focus Point 10.3). While the physical activity component of these games is relatively basic at this point, it is likely to become increasingly more sophisticated as the virtual reality aspect of computer technology advances. Developments in broadcast technology have seen the emergence of off-site 'fan parks' where big screen broadcasts of the sporting events may actually rival spectator numbers at the site of the competition. This trend has led Weed (2010) to question whether basic assumptions that have been traditionally held related to sport travel motivations are still valid. Buckley *et al.* (2015) predict that tourists will increasingly take on the traits of cyborgs as our physical and sensory capabilities are augmented through built devices. This is already happening through tourism-related smartphone applications and wearable aids like Google Glass and Samsung Gear headsets that enable viewers to control camera angles when watching a sport broadcast.

Focus Point 10.3: The E-Sport Revolution

While the conceptualisation of sport tourism in this book clearly identifies gross motor movement as a key characteristic of sport, this view is being challenged with the emergence of e-sports. Chen (2017) reports that 43 million people tuned into a multiplayer competition of League of Legends that was held in various cities throughout North America in 2017. While many of these viewers watched from the screens of their laptops, iPads or smartphones, it is likely that many travelled to watch or participate in the competition in person. Indeed, the combatants in these sports can compete with each other (in teams) or against each other (individual or team opponents) in real time from anywhere in the world with ultrafast broadband. These types of gaming competitions are becoming increasingly popular and offer substantial earning potential for the top players. A variety of gaming themes have emerged including digital versions of real-life sports like EPL football and NBA basketball. 'Real' sport leagues have used these games to promote their brand and to win new or retain existing markets. China has proved to be a fertile ground for e-sports with the national State General Administration of Sport establishing a national e-sports team. Further evidence of the growing acceptance of e-sports is evident in the fact that they will be included in the sporting programme of the 2022 Asian Games in Hangzhou, China. Regardless of one's personal view of whether e-sports classify as real sport or not, it is clear that if this trend continues it will have a growing impact on sport and on sport tourism. As gaming technology advances, the e-sport landscape will continue to evolve rapidly changing the world of sport and tourism, and the experience of sport tourism as we know it.

Key Reading

Chen, S. (2017) Big crowds, big bucks: A beginner's guide to the e-sports phenomenon that's conquering China. *South China Morning Post.* See http://www.scmp.com/news/china/society/article/2104977/big-crowds-big-bucks-beginners-guide-e-sports-phenomenon-thats.

Past technological advances in transportation have played a key role in tourism development and they are likely to continue to do so. For example, the air transport industry is under substantial pressure to reduce its carbon footprint. Futurologist Robin Manning notes several trends that will help the industry to improve in this area:

Research in nanotechnology, biotechnology, information technology and cognitive science are providing a growing set of opportunities. Some examples include; new light and strong composite materials, electronic plastics, fuels created directly from growing plants (that are effectively scrubbing the atmosphere of unwanted carbon), smarter computer and avionics systems and transport informatics. (Nusca, 2010: np)

Above the earth's atmosphere, continued advances in space travel will result in increased access to weightless environments, which may spawn a whole new generation of sport activities. Similar developments in marine environments are likely to present dramatic new opportunities for sport tourism in that realm.

Conclusion

Sport and tourism are each dynamic in and of themselves. While traditional patriot games like the Olympics and the FIFA World Cup and the traditional competitive sports that they represent continue to grow, they do so at a slower rate than less competitive individual adventure and lifestyle sports. In a tourism context, there have also been many changes with one of the more substantial ones being a noticeable shift towards an experience-based economy. The interaction of these types of trends across both realms has made sport tourism an extremely innovative and exciting environment in which to play, work and study. It is evident that sport influences the nature and pace of tourism destination life cycles and that tourism has a similar impact on sport life cycles. The conscious manipulation of these forces offers a powerful tool for pursuing sustainable development strategies.

Heritage sport tourism represents another major dynamic in the temporal dimension of sport tourism, which has driven tourists to seek out their idealised pasts by visiting places infused with sport heritage and by participating in programmes that bring the past to life. The sport tourism industry has only recently started to appreciate the breadth of products that are of interest to these tourists. Heritage sport tourism offers the opportunity for sport tourists to become time travellers if only in their imaginations. Rather than having a myopic focus on the geographic dimension of sport tourism, researchers and managers need to recognise the temporal dimension of sport tourism experiences as it expands the range of possible sport tourism experiences exponentially.

It is also clear that sport tourism does not operate in a vacuum. There are a variety of trends in the economic, environmental, political, socio-demographic and technological realms that may have a direct and, in some cases, a dramatic influence on sport tourism. Other impacts may be indirect and less dramatic but when aggregated over time they may be substantial and game changing. Climate change is a case in point. On a year-to-year basis, the incremental impacts of climate change are often lost in the normal variations of seasonality and weather. However, in the long term, climate change will have a major impact on the spatial and temporal distribution of sport and tourism at the global and regional levels. By studying trends that exist within the external environment, sport tourism managers will be in a better position to set sustainable sport tourism development goals and objectives, and to develop effective plans of action. The pursuit of sustainable development through sport-related tourism requires that these trends are recognised, understood and acted upon.

Part 5
Conclusions

11 Shifting Goal Posts and Moving Targets: The Ever-Evolving Worlds of Sport and Tourism

Introduction

The pervasiveness of sport in society is universal. Indeed, the International Olympic Committee claims sport to be a human right, and the United Nations (2017) considers sport to be a low-cost, high impact tool for humanitarianism and human development. It might be argued that sport shapes the lives of individuals and the societies in which individuals live, but it is equally true that sport is a reflection of society. Sport is, like tourism, a socially constructed phenomenon (Andrews, 2007). Although sports can be defined in technical terms, sport phenomena are a reflection of their spatio-historical context. The sports that Alexander the Great hosted on the eve of his Persian campaign in 334 BC, and the symbolism and imagery that he sought to embellish through the hosting of the Olympian Games, were unique to those Classical times. Equally, today, contemporary practices of sport as they relate to meanings, identity and lifestyles, are a unique and diverse reflection of societies that are ever-evolving.

In this respect, we can reflect upon rapid changes in our own societies that have unfolded since the publication of the two previous editions of this book; the growth in sub-elite engagement in sports events by 'serious leisure' competitors, the measurement of personal performance that is self-referenced rather than results oriented, the growth of online gaming and virtual sports, the continuing evolution of single sports into hybrid forms and, perhaps most notably, the phenomenal growth and diversity of individualised, unstructured, freestyle sports (Andrews, 2006; Coakley, 2007). These new generation sports have strong subcultural associations, and offer participants enriching opportunities to build a sense of personal or collective identity (Hagen & Boyes, 2016), in which performance and results may be completely unimportant, or at the very least secondary to aesthetics, style and meaning (Falcous, 2017). These dynamics of sport are a reflection of individual identities that have been confused, and national identities that have been eroded by the forces of globalisation (Higham & Hinch, 2010).

Within this socially dynamic context, our first contribution in this book is to consider the defining criteria of sport, and how these criteria have evolved in relation to contemporary practices of sport and tourism. Technical definitions of sport address those defining criteria that are universal, beyond dispute and immutable across space and time. In this respect, we have long used the definition of McPherson *et al.* (1989: 15) who state that sport is 'a structured, goal-oriented, competitive, contest-based, ludic physical activity'. Of course, these immutable technical criteria are open to interpretation, and as they are reinterpreted so the vast diversity of sports and forms of engagement in sports arise (Hinch & Higham, 2004). So, while sport is structured by rules of engagement, the implementation of rules varies from the strictly enforced to the unwritten, unspoken or deliberately ignored and unenforced. Rules may be codified to be learned and periodically revised; or may be unspoken and available only to members of a sport subculture (Wheaton, 2004).

McPherson *et al.* (1989) go on to address the goal orientation of sport, noting that participants seek to attain certain levels of achievement or competence but, once again, goal orientation is open to subjective interpretation. Goal orientation may be commonly expressed in terms of victory over, or defeat at the hands of, an opponent, but this is a view of goal orientation that is incomplete without consideration of ego (being the best) and task (doing my best) orientations (Duda & Nicholls, 1992) such that goal orientation may be result focused or self-referenced (Falcous, 2017). These are critically important points of distinction in relation to tourism experiences. Competition is typically measured in performance against an individual opponent or rival team, but extends to competing against personal standards, degrees of difficulty, virtual opponents, the forces of nature or a technically challenging course. The uncertainty of outcomes associated with competition ensures that at its best, sport is unrivalled in terms of atmosphere, the potential for drama and sustained audience engagement.

While movement is a requirement of physical activity, spatial mobility is a requirement of sports as spectators and participants search for appropriate, desirable or iconic locations and venues of sport engagement and competition. These are defining qualities of sport (Hinch & Higham, 2004), and the ever-evolving interpretations of technical elements of sport make for a dynamic and fascinating field of research. While we address this field of research in chapters that may be read as a series of individual essays on sport-related tourism, collectively the preceding chapters were integrated by a conceptual framework that gives this book its structure. The dynamic spatio-temporal constructions of sport and tourism lead us to consider contemporary practices of sport and tourism in relation to space and time. Following the introduction (Chapter 1), we presented three framing chapters (Chapters 2–4) that addressed conceptual foundations for the study of sport and tourism (Chapter 2), sport tourism markets (Chapter 3) and development processes for sustainable sport tourism (Chapter 4). These framing chapters served as a foundation for the chapters that followed which were organised into two major parts. Chapters 5–7 addressed the spatial dimensions of sport tourism, which included the geographical concepts of space (location and travel flows), place (sport and culture) and environment (landscape, resources and impacts), respectively. Chapters 8–10 then addressed the temporal dimensions of sport tourism by considering the short-term (sport and the tourist experience), medium-term (seasonality, sport and tourism) and long-term (evolutionary trends in sport tourism) temporal frames, respectively.

The study of sport tourism development provides varied and important insights into the ways in which this phenomenon has changed, and continues to change, across space and time. Chapter 1 raised three fundamental questions: 'What makes sport unique as a focus for tourism development?', 'How is sport tourism manifest in space?' and 'How do these manifestations change over time?'. In addressing these questions, an objective of this book was to capture and critically analyse the increasingly diverse manifestations of sport tourism development. Expanding personal mobilities and increasingly diverse engagements in sport are such that the intersection of the two has become an area of considerable researcher interest (Gibson, 2004; Glyptis, 1982; Maguire, 1993, 1994; Standeven & De Knop, 1999; Weed & Bull, 2004). There has been an enduring focus of scholars on sports mega-events (Getz, 1997; Ritchie, 1984), but it has been our longstanding contention that sport tourism phenomena not only include but also extend far beyond spectator events. Indeed, Cornelissen (2004: 40) notes 'since the vast majority of mega-events are hosted by industrialized states, discourse and research on the processes and impacts of these events tend to be framed around the economic and political circumstances characteristic to the developed world'. In this book, we draw attention to the wider manifestations of sport, and the need for critical scholarly work that embraces the broad diversity of sport-related tourism.

Foundations for Sport Tourism Development

The study of sport tourism development requires a foundation that includes an underlying framework highlighting the relationship between sport and tourism, an appreciation of sport tourism markets and an understanding of fundamental development concepts and issues. The opening chapter of Part 2 of the book conceptualises sport tourism as sport-based travel away from the home environment for a limited time where sport is characterised by unique rule sets, competition related to physical prowess and a playful nature. From this perspective, sport is viewed as a tourist attraction based on an updated version of Leiper's (1990) tourist attraction system that articulates the major categories of sport tourists and a fourfold typology of the sport attraction nucleus (spectator events, participant events, active engagement in recreational sport and sport heritage). By consciously treating sport as a unique type of tourist attraction, readers will be in a better position to understand the nature of and to influence patterns of sport tourism development. This perspective, as well as alternative perspectives, is increasingly based on a substantive body of literature and academic offerings (e.g. Gibson, 2006; Weed, 2006, 2009b). Case Study 2.1 (Richard Shipway) illustrated the progress achieved in the study of sport tourism as manifest in Bournemouth University's sport tourism curriculum and its strategy of embedding, enthusing and enhancing employability for students in this field.

A critical appreciation of the nature of sport tourism markets is also of foundational importance to the study of sport tourism development. Sport tourism is not only a specialised segment of the tourism market, but it is also comprised of multiple niche markets. The nature of these niche markets varies with different forms of engagement in sport, with implications for tourist motivations and experiences. In this book, we move beyond the tripartite classifications of sport-related tourism that have been previously proposed by Redmond (1990) and Gibson (1998a), who consider sport tourism to exist in the form of sport vacations (active), multisport festivals and world championships (events), and sports halls of fame and museums (nostalgia). We extend our discussions by way of a fourfold classification of sport tourism that includes spectator events, participation events, active engagement in recreation sports and sports heritage.

In Chapter 3, we considered leisure sports markets, and in Case Study 3.1, Eiji Ito examined the relationship between culture, ideal affect and sport tourist motivation. Specifically, this classification also allowed for separate treatment of spectator and participation events and our justification for this lies in the phenomenal growth in recent years, and the increasing diversity that now exists in participation sports events. We consider the development of participation events to be a feature of the evolution of sport tourism since the publication of the previous edition of this book in 2010. Many such participants engage in serious training schedules in preparation for their target events, but for most their actual performance is self-referenced, as opposed to results orientated (Falcous, 2017). For others, engagement in participation sports events may be considered 'de-sportified' in that they are 'loosely structured, non-competitive, and socially connective' (Falcous, 2017: 1). This schema underpins our discussions in Chapter 3: Sport Tourism Markets and later in Chapter 8: Sport and the Tourist Experience.

Sport tourism development implies progress that may be measured in terms that extend beyond economic growth to personal growth and fulfilment, improved quality of life and various other indicators of personal and social well-being. Planning is required to guide and direct development processes to obtain desired future states. In keeping with the UN's goals for sustainable development, long-term perspectives and active interventions are needed if sport tourism development is to be sustainable. Three key development issues were explored in this book. The first relates to commodification and authenticity. Although tourism is just one of the forces that is commodifying sport, it is recognised as an important one. A key advantage of sport as a tourist attraction is its propensity for authenticity. This is found in its uncertain outcomes, displayed as part of performance; its physical basis and all-sensory nature; self-making and the construction of identity; and its ability to foster community. Sport provides visitors with access to the 'backstage' of a destination in which residents, whether spectators or athletes, cease to perform for the visitors. In this sense, sport offers a portal into the 'real' community.

Globalisation is the second key development issue. Sport is a high profile manifestation of globalisation featuring professional leagues that increasingly cross borders and recruit players from a global pool. Sports such as football (soccer) are global phenomena yet there remain important distinctions in the way the game is played in different parts of the world. These differences represent opportunities for sport tourism development. In such cases, sport can contribute to the unique regional character of a place thereby providing a competitive tourism advantage. Major investments in sport events, like South Africa's hosting of the 2010 FIFA World Cup, are justified as ways to position cities and countries in the global marketplace (Case Study 4.1, Brendon Knott). The third major development issue highlighted was the challenge of organisational fragmentation. Attempts to overcome the problem of fragmentation through the creation of sport tourism alliances have been encouraging. Continued articulation of the benefits of cooperation is required. If these benefits cannot be demonstrated, it is unlikely that the various stakeholders in sport tourism, including sports event organisers and promoters, sports associations, managers of sporting venues, destination managers and tourism marketers, will work in a cooperative fashion.

Sport Tourism Development and Space

Key themes within the spatial analysis of sport tourism development are the study of the locations and travel flows associated with sports; the way that sport infuses space with meaning to create unique tourism places; and the resource requirements and impacts of sport tourism. In Chapter 5, we specifically addressed the spatial parameters of sport tourism, drawing on the geographical concepts of systems and scale, central place theory, location hierarchies, spatial travel flows and centre and periphery. Aspects of sport, tourism and space were critically illustrated in Case Study 5.1 (Arianne Reis) in relation to the Rio de Janeiro (Brazil) 2016 Olympic Games and the geographies of exclusion; and in Focus Points 5.1 and 5.2 in relation to stadium location hierarchy and location flux, respectively. We find these spatial geographical tools particularly useful in explaining the locational tendencies of urban sport facilities, events and professional sport teams. Sport tourism in peripheral areas is characterised by quite different spatial criteria.

Place was described in this book as space infused with meaning. It is especially attractive to tourism marketers who use sport to sell destinations. Hosts and visitors develop attachment to these places based on a combination of their dependence on the attributes found at the place and the extent to which the place contributes to their identity. Case Study 6.1 (Daniel Evans) illustrated the tensions that can arise through the interplay of the local with the global. In this case, the place identity of local supporters of Liverpool's football team was being challenged by the presence of an international fan base. Many of these international fans make the pilgrimage to Liverpool to support 'their' team despite their lack of residence in the city. In a similar manner, the spectre of homogenised sportscapes threatens tourism, as it contributes to a breakdown of distinctive regional place meaning based on sport. A challenge therefore exists to retain the unique regional meanings and identities associated with sporting facilities and activities while participating in the global marketplace. Tourism interests should support the efforts of sport managers who defend the 'integrity' of their sports. There should also be consistency between the images used to promote a sport tourism destination and the ways that residents identify their home. Conflicting views of place may result in conflicting attitudes and behaviours.

The environment and, more specifically, landscapes, resources and impacts were the focus of the final chapter under the spatial theme (Chapter 7). Here, we drew upon Bale's (1989) claim that sports are a cultural form, and that as such the landscapes of sport are part of the cultural landscape. Thus, sports that utilise elements of the natural environment may subject those landscapes to varying degrees of anthropogenic change. In this chapter, we considered the resource base for sport and tourism, and discussed the importance of sustainable management of those resources. The pressing challenges arising from the globally dispersed impacts of sport tourism were highlighted. Case Study 7.1 (Debbie Hopkins) specifically considered the environmental ironies that are embodied in the sport-related mobilities associated with winter sport destinations and the decline in the natural resources – specifically snow – upon which winter sports resorts so depend. These challenges were reflected in Focus Point 7.1 (Michelle Rutty), which highlighted the water-intensive nature of the sport of golf. This chapter also considered sport tourism in the built environment, contemplating issues associated with the impacts of sports facility development in relation to the concepts of landscape and sportscape.

Sport Tourism Development and Time

The temporal dimensions of sport tourism include tourist experiences of sport, seasonal variations in sport and tourism and the long-term evolutionary dynamics of sport tourism phenomena. There are two noteworthy features of Chapter 8. First, our considerations of the sport tourism experience was informed by Morgan's (2007) experience space model, which presents two contrasting approaches to the delivery of sports experiences. The managerial approach, as described by Morgan (2007), approaches the experience as a product or service to add experiential value to the sporting competition in a way that is predetermined and standardised. The Olympic Opening and Closing ceremonies, which are choreographed, scripted and rehearsed, are examples of this approach. The alternative consumer behaviour approach recognises that the experiences of the sport tourist may be emotional, symbolic and transformational, if the personal, social and cultural meanings of the event for both hosts and guests can be fostered to allow for the co-creation of sport experiences based on intense social interaction. The importance that consideration is given to the merits of these approaches was illustrated by way of a discussion of Japan's planning for the 2019 Rugby World Cup (Focus Point 8.1). Secondly, our fourfold classification of sport tourism, which includes spectator events, participation events, active engagement in recreational sports and sports heritage, gave us the structure of this chapter. The growth of participation sports events was also a feature of this chapter and, in this respect, Case Study 8.1 (Moyle, Kennelly and Lamont) offered insights into the pursuit of serious sporting engagements by amateur athletes, in relation to both their own experiences and those of their spouses and entourages.

Seasonal patterns in sport and tourism represent a distinct temporal dimension of sport tourism development. Tourism managers have typically seen seasonal variation in visitors as a problem due to underutilised capacity and decreased revenue flows during low season. From a tourism management perspective, sport can be used as a strategy to influence seasonality with considerable success. The scheduling of sporting events during the shoulder tourism seasons is an increasingly prominent aspect of event production and planning. In terms of active sport tourism, destinations that are characterised by specialised resources available during off-season can be particularly attractive to sport subculture groups. Similarly, nostalgia sport tourism attractions remain functional outside the main tourism seasons. If properly promoted, they too offer the opportunity to alter seasonal patterns of visitation to a destination. More recently, climate change is having a significant impact on the seasonality of sport tourism. This is particularly true for sport activity that is directly tied to climate conditions like skiing (Case Study 9.1, Robert Steiger) which faces the general challenge of shortened 'natural' ski seasons although this varies by location.

Tourism destination cycles as well as sport life cycles offer valuable theoretical insight into the long-term evolution of sport and tourism (Chapter 10). As sports progress through their life cycles, they impact the destinations where they take place. For example, the shift from traditional competitive sport to individual and lifestyle sports will impact the distribution of sport tourism development as new sites are developed based on new resources needs. Case Study 10.1 (Adam Doering) provides a good example with its articulation of the way that surfing has emerged as both a sport and a factor in the distribution of sport tourism. Individual sport life cycles are also influenced by tourism. As leisure-oriented activities like surfing grow as popular tourist activities, variations have emerged including high-performance competitions such as the inclusion of surfing in the 2020 Tokyo Olympics.

Heritage sport tourism demonstrates how the evolution of sport can have a direct impact on tourism. Sport nostalgia is, in part, a reaction to a rapidly changing society. It is not only a chance for tourists to revisit their youth, but it is also an opportunity to escape back into what is often considered a simpler time. These temporal journeys are facilitated by spatial journeys as tourists' seek out sports halls of fame, sites of historic sporting moments and fantasy programmes that enable them to recreate or relive history. More generally, there are a variety of macro-level trends that will influence the future of sport tourism. Globalisation processes appear to underlie the majority of these trends. Dynamic economic, environmental, political, socio-demographic and technological/communication realms present opportunities as well as challenges for sport tourism development. Not only is each of these realms characterised by changing characteristics, but these characteristics interact with those found in the other realms, making cause/effect relationships difficult to identify, let alone predict. As difficult as this exercise may be, the ability to influence sport tourism development depends in part on being able to do so.

The Final Whistle

Sport provides a unique, dynamic and intriguing focus for the study of tourism. Engagement in sports in societies is extremely wide ranging, and engagement in sport-related tourism is a dynamic phenomenon. The significant profile of sports in media markets, ranging from the local to the global, presents interesting interplays of sport, media and tourism interests. Perhaps more importantly, the fundamental rule structures, competitive dimensions and ludic qualities of sport present a complex array of opportunities and challenges associated with tourism development. Thus, sport is characterised by markets that are distinguishable from those associated with other forms of tourism and development issues that warrant dedicated research and publication.

This book explored the functions of sport as a distinctive and potentially powerful type of tourist attraction. The defining qualities of sport provide an exceptional and unique attraction nucleus that offers fertile grounds for tourism industry interests and sport tourism researchers to explore. The human element of sport as a tourist attraction is notable for its diversity. Among the many dimensions or continua upon which sport tourists may be conceptualised, elite athletes and recreational participants, event competitors and sport spectators, sports teams and individual contestants, self-referenced competitors and participants in recreational sports are just a few. Ever-evolving forms of sport tourism may be explored in terms of diverse market niches, desired visitor experiences and (social) media reach, which mark destinations as unique.

In treating sport as a peculiar and distinctive type of tourist attraction, this book demonstrates that sport tourism has real manifestations in space and time. Sport tourism influences travel patterns, sport and tourism locations, the meanings ascribed to tourism destinations and impacts on the landscapes, built (e.g. stadiums and arenas) and natural (e.g. marine environments and ski areas), which serve as theatres of sport. It also influences the nature of travel experiences, seasonal visitation patterns and the evolution of sport, tourism and the host destinations where sport tourism takes place. While it is important to recognise changing patterns of sport-related tourism, it is also important to critically understand the drivers of dynamic change in ways that are theoretically informed. Research in sport tourism has moved beyond description to the realms of explanation and prediction as facilitated by theory.

At this stage in the evolution of the study of sport tourism, theoretical insights have tended to come from advanced fields of scholarship. The theoretical insights that we have presented in this book have been drawn primarily from geography. To the extent that these theories provide additional insight into sport-related tourism phenomena, they should be adopted, applied and further developed in their application to the study of sport tourism. The driving forces behind such theory building should include not only the obvious need to solve real and potential problems related to the practice of sport tourism but also curiosity-driven research with its less targeted outcomes, but with the potential to provide new insights into the phenomenon. An appreciation of these dynamics means that stakeholders do not have to be passive observers or reactive players in the sport tourism development process. Active engagement in theory building and empirical research will continue to advance our understanding of the dynamic relationship between sport and tourism.

We opened this book by recounting some of the functions of sport in Alexander the Great's Macedonian society in 334 BC. It therefore seems appropriate to close this chapter with some reflections upon our current societies, which frame contemporary practices of sport and tourism today. It has been widely recognised that sport and tourism are socially constructed phenomena that should be founded on principles of opportunity and inclusion. The United Nations (2017) considers sport to be a fundamental human right with physical activity central to human health, while tourism is increasingly viewed in relation to quality of life and subjective well-being (McCabe, 2009; McCabe & Johnson, 2013; McCabe et al., 2010), as well as in relation to social and family capital (Minnaert et al., 2009). Yet, participation in many sports and recreational pursuits remains rigidly defined by social class, race and gender (Gibson, 2005). 'Irrespective of culture or historical period, people use sport to distinguish themselves and to reflect their status and prestige' (Booth & Loy, 1999: 1). Participation in physical activities remains beset with concerns surrounding increasingly sedentary and unhealthy lifestyles and obesity in some societies (Coakley, 2007). Exclusion continues to be a feature of many contemporary sports. Simultaneously, many children who are too young to make their own decisions are vulnerable to exploitation by over-ambitious parents, results-driven coaching regimes and intensely competitive national sport development programmes.

Booth and Loy (1999) state that similar status groups generally share lifestyle and consumption patterns, and the same is true of tourism practices where, despite concerns about the energy demands and emissions of tourist transportation (Becken, 2007; Buckley et al., 2015), frequency of travel and distance continue to be used to distinguish between social strata and entrenched middle-class social relationships (Cohen, 1984). It has been widely recognised that high personal mobility (Burns & Novelli, 2008; Hall, 2004) and increasing interest in both passive and active engagements in physical activities (Glyptis, 1991) are two defining features of late 20th-century and early 21st-century societies, yet in many societies personal mobility and freedoms to travel are constrained and exclusive (Hall, 2004). This leads us to conclude that governments must, if they are not already doing so, develop national and regional policies (and data, measurement and reporting systems) that align with the United Nations (2015) Sustainable Development Goals (SDGs), which embody a range of critical contemporary issues relating to the transformation of societies in the interests of sustainability. It is clear that social inclusion remains an issue in both sport and tourism; addressing social inclusion relates directly to SDGs 3 (good health and well-being), 5 (gender equity), 10 (reduced inequalities), 12 (responsible consumption and production), 16 (promote just, peaceful and inclusive societies) and 17 (partnerships for sustainable development). Equally, it is important to be acutely aware that our societies face immense challenges relating to sustainable development, including rising inequalities, global health threats, environmental degradation, natural resource depletion and climate change. The social and political forces that shape our sport and tourism practices have critical roles to play in transforming our societies to meet these challenges.

References

Abad, J.M. (2001) The growth of the Olympic City of Barcelona. *Olympic Review* 27 (38), 16–19.

Abegg, B. and Frösch, R. (1994) Climate change and winter tourism. Impact on transport companies in the Swiss canton of Graubünden. In M. Beniston (ed.) *Mountain Environments in Changing Climates* (pp. 328–340). London/New York: Routledge.

Abramson, A. and Fletcher, R. (2007) Recreating the vertical: Rock-climbing as epic and deep eco-play. *Anthropology Today* 23 (6), 3–7.

Ajzen, I. and Driver, B.L. (1992) Application of the theory of planned behavior to leisure choice. *Journal of Leisure Research* 24 (3), 207–225.

Alexander, M. (2001, 15 April) Sport relays billions to NZ economy. *Sunday Star Times*, p. E:3.

Alexandris, K. and Carroll, B. (1999) Constraints on recreational sport participation in adults in Greece: Implications for providing and managing sport services. *Journal of Sport Management* 13 (4), 317–332.

Allcock, J.B. (1989) Seasonality. In S.F. Witt and L. Moutinho (eds) *Tourism Marketing and Management Handbook* (pp. 387–392). Englewood Cliffs, NJ: Prentice Hall.

Allen Collinson, J. and Hockey, J. (2007) 'Working out' identity: Distance runners and the management of disrupted identity. *Leisure Studies* 26, 381–398.

Allmers, S. and Maennig, W. (2009) Economic impacts of the FIFA Soccer World Cups in France 1998, Germany 2006, and outlook for South Arica 2010. *Eastern Economic Journal* 35, 500–519.

Amelung, B., Nicholls, S. and Viner, D. (2007) Implications of global climate change for tourism flows and seasonality. *Journal of Travel Research* 45, 285–296.

Andrews, D. (2006) *Sports-Commerce-Culture: Essays on Sport in Late Capitalist America*. New York: Peter Lang.

Andueza, J.M. (1997) The role of sport in the tourism destinations chosen by tourists visiting Spain. *Journal of Sport Tourism* 4 (3), 5–12.

Anon. (1996, 18 July) Business money is the champion. *Marketing* 3.

Anon. (2000) Shared vision? Australian leisure management considers the development of shared community and school sports facilities in country areas. *Australian Leisure Management* 19, 24–25.

Archer, B. and Cooper, C. (1994) The positive and negative impacts of tourism. In W. Theobald (ed.) *Global Tourism: The Next Decade* (pp. 73–91). Oxford: Butterworth Heinemann.

ASICS (nd) ASICS Sports Museum. See http://corp.asics.com/en/about_asics/museum (accessed 22 August 2017).

Atkisson, A. (2000) *Believing Cassandra: An Optimist Looks at a Pessimists World*. New York: Scribe Publishers.

Bagheri, A. and Hjorth, P. (2007) Planning for sustainable development: A paradigm shift towards a process-based approach. *Sustainable Development* 16, 83–96.

Baker, C. (2015) Beyond the island story?: The opening ceremony of the London 2012 Olympic Games as public history. *Rethinking History* 19 (3), 409–428.

Bale, J. (1982) *Sport and Place: A Geography of Sport in England, Scotland and Wales.* London: C. Hurst & Co. Ltd.

Bale, J. (1989) *Sports Geography.* London: E&FN Spon.

Bale, J. (1993) The spatial development of the modern stadium. *International Review for the Sociology of Sport* 28 (2–3), 121–133.

Bale, J. (1993a) *Sport, Space and the City.* London: Routledge.

Bale, J. (1993b) The spatial development of the modern stadium. *International Review for the Sociology of Sport* 28 (2+3), 121–134.

Bale, J. (1994) *Landscapes of Modern Sport.* Leicester: Leicester University Press.

Bale, J. (2002) *Sports Geography* (2nd edn). London: Routledge.

Bale, J. and Maguire, J. (2013) *The Global Sports Arena: Athletic Talent Migration in an Interdependent World.* London: Routledge.

Ball, R.M. (1988) Seasonality: A problem for workers in the tourism labour market? *The Services Industries Journal* 8 (4), 501–513.

Ball, R.M. (1989) Some aspects of tourism, seasonality, and local labour markets. *Area* 21 (1), 35–45.

Baloglu, S. and McCleary, K.W. (1999) A model of destination image formation. *Annals of Tourism Research* 26 (4), 868–897.

Barker, M. (2004) Crime ans sport events tourism: The 1999/2000 Americas Cup. In B. Ritchie and D. Adair (eds) *Sport Tourism: Interrelationships, Impacts and Issues* (pp. 226–252). Clevedon: Channel View Publications.

Barney, R.K., Wenn, S.R. and Martyn, S.G. (2002) *The International Olympic Committee and the Rise of Olympic Commercialism.* Salt Lake City, UT: University of Utah Press.

Baron, R.R.V. (1975) *Seasonality in Tourism: A Guide to the Analysis of Seasonality and Trends for Policy Making.* London: Economist Intelligence Unit.

Barros, C.P., Butler, R. and Correia, A. (2010) The length of stay of golf tourism: A survival analysis. *Tourism Management* 31 (1), 13–21.

Bartoluci, M. and Čavlek, N. (2000) The economic basis of the development of golf in Croatian tourism: Prospects and misconceptions. *Acta Turistica* 12 (2), 105–138.

Basińska-Zych, A. and Lubowiecki-Vikuk, A.P. (2011) Sport and tourism as elements of place branding. A case study on Poland. *Journal of Tourism Challenges and Trends* 4 (2), 33–52.

Basset, C. and Wilbert, C. (1999) Where you want to go today (like it or not). In D. Crouch (ed.) *Leisure/Tourism Geographies* (pp. 181–194). London: Routledge.

Baum, T. and Hagen, L. (1999) Responses to seasonality: The experiences of peripheral destinations. *International Journal of Tourism Research* 1, 299–312.

Baum, T. and Lundtorp, S. (2001) Seasonality in tourism: An introduction. In T. Baum and S. Lundtorp (eds) *Seasonality in Tourism* (pp. 1–4). London: Pergamon.

Beardsley, D. (1987) *Country On Ice.* Markham, Ontario: Paperjacks Ltd.

Beezer, A. and Hebdige, D. (1992) Subculture: The meaning of style. In M. Barker and A. Beezer (eds) *Reading Into Cultural Studies* (pp. 101–117). London: Routledge.

Befu, H. (2001) *Hegemony of Homogeneity: An Anthropological Analysis of 'Nihonjinron'.* Melbourne: Trans Pacific Press.

Belanger, A. (2000) Sport venues and the spectacularization of urban spaces in North America. *International Review for the Sociology of Sport* 35 (3), 278–397.

Bell, R. (2000) A modern perspective of the ancient Olympic events. *The Sport Journal* 3 (3), 1–2.

Bellan, G.L. and Bellan-Santini, D.R. (2001) A review of littoral tourism, sport and leisure activities: Consequences on marine flora and fauna. *Aquatic Conservation: Marine and Freshwater Ecosystems* 11 (4), 325–333.

Bentley, T.A., Page, S.J. and Laird, I.S. (2000) Safety in New Zealand's adventure tourism industry: The client accident experience of adventure tourism operators. *Journal of Travel Medicine* 7 (5), 239–245.

Beresford, S. (1999) The sport–tourism link in the Yorkshire region. In M. Scarrott (ed.) *Exploring Sports Tourism: Proceedings of a SPRIG Seminar Held at the University of Sheffield on 15 April 1999* (pp. 29–37) Sheffield: Sheffield Hallam University.

Bernstein, A. (2000) Things you can see from there you can't see from here: Globalization, media, and the Olympics. *Journal of Sport and Social Issues* 24 (4), 351–369.

Bieger, T. and Laesser, C. (2002) Market segmentation by motivation: The case of Switzerland. *Journal of Travel Research* 41 (1), 68–76.

Binns, T. (1995) Geography in development: Development in geography. *Geography* 80, 303–322.

Black, D. (2008) Dreaming big: The pursuit of 'second order' games as a strategic response to globalization. *Sport in Society* 11 (4), 467–480.

Bloch, C. and Laursen, P.F. (1996) Play, sports and environment. *International Review for the Sociology of Sport* 31 (2), 205–217.

Bodet, G. and Lacassagne, M.F. (2012) International place branding through sporting events: A British perspective of the 2008 Beijing Olympics. *European Sport Management Quarterly* 12 (4), 357–374.

Boniface, B.G. and Cooper, C. (1994) *The Geography of Travel and Tourism* (2nd edn). Oxford: Butterworth Heinemann.

Boorstin, D.J. (1975) *The Image: A Guide to Pseudo-Events in America.* New York: Atheum.

Booth, D. (1993) Sydney 2000: The games people play. *Current Affairs Bulletin* December/January, 4–11.

Booth, D. (1997) Sports history: What can be done? *Sport, Education and Society* 2 (2), 191–204.

Booth, D. (1999) Gifts of corruption? Ambiguities of obligation in the Olympic movement. *Olympika* 8, 43–68.

Booth, D. (2000) Modern sport: Emergence and experiences. In C. Collins (ed.) *Sport in New Zealand Society* (pp. 45–63). Palmerston North, New Zealand: Dunmore Press.

Booth, D. and Loy, J.W. (1999) Sport, status, and style. *Sport History Review* 30, 1–26.

Borish, L.J. and Rischler, B.L. (2001) Labour, leisure and sport in cultural perspective. *Rethinking History* 5 (1), 1–9.

Borland, J. and MacDonald, R. (2003) Demand for sport. *Oxford Review of Economic Policy* 19 (4), 478–502.

Bourdeau, P., Corneloup, J. and Mao, P. (2002) Adventure sports and tourism in the French Mountains: Dynamics of change and challenges for sustainable development. *Current Issues in Tourism* 5 (1), 22–32.

Bows-Larkin, A., Mander, S.L., Traut, M.B., Anderson, K.L. and Wood, F.R. (2016) Aviation and climate change: The continuing challenge. *Encyclopedia of Aerospace Engineering.* doi: 10.1002/9780470686652.eae1031.

Boyle, R. and Haynes, R. (1996) 'The grand old game': Football, media and identity in Scotland. *Media, Culture & Society* 18 (4), 549–564.

Bramwell, B. (1999) Sport, tourism and city development. *International Journal of Tourism Research* 1 (6), 459–460.

Bramwell, B., Higham, J.E.S., Lane, B. and Miller, G. (2017) Twenty-five years of sustainable tourism: Looking back and moving forward. *Journal of Sustainable Tourism* 25 (1), 1–9.

Brayley, R.E. (1999) Using technology to enhance the recreation education classroom. *Journal of Physical Education, Recreation & Dance* 70 (9), 23–25.

Breivik, G. (2010) Trends in adventure sports in a post-modern society. *Sport in Society* 13 (2), 260–273.

Breuer, C., Hallmann, K. and Wicker, P. (2011) Determinants of sport participation in different sports. *Managing Leisure* 16 (4), 269–286.

Briassoulis, H. (2007) Golf-centered development in coastal Mediterranean Europe: A soft sustainability test. *Journal of Sustainable Tourism* 15, 441–462.

Briassoulis, H. (2010) 'Sorry golfers, this is not your spot!': Exploring public opposition to golf development. *Journal of Sport and Social Issues* 34 (3), 288–311.

British Tourist Authority (2000) *Sporting Britain: Play It, Love It, Watch It, Live It, Visit.* London: Haymarket Magazines Ltd.

Broudehoux, A-M. (2016) Mega-events, urban image construction, and the politics of exclusion. In R. Gruneau and J. Horne (eds) *Mega-Events and Globalization: Capital and Spectacle in a Changing World Order* (pp. 113–130). London: Routledge.

Brown, C. and Paul, D.M. (1999) Local organized interests and the 1996 Cincinnati Sports Stadia Tax Referendum. *Journal of Sport and Social Issues* 23 (2), 218–237.

Brown, G. (2000) Emerging issues in Olympic sponsorship: Implications for host cities. *Sport Management Review* 3 (1), 71–92.

Brown, G. (2001) Sydney 2000: An invitation to the world. *Olympic Review* 27 (37), 15–20.

Brown, G. and Raymond, C. (2007) The relationship between place attachment and landscape values: Toward mapping place attachment. *Applied Geography* 27 (1), 89–111.

Brown, G., Chalip, L., Jago, L. and Mules, T. (2002) The Sydney Olympics and brand Australia. In N. Morgan, A. Pritchard and R. Pride (eds) *Destination Branding: Creating the Unique Destination Proposition* (pp. 163–185). Oxford: Butterworth-Heinemann.

Brown, G., Smith, A. and Assaker, G. (2016) Revisiting the host city: An empirical examination of sport involvement, place attachment, event satisfaction and spectator intentions at the London Olympics. *Tourism Management* 55, 160–172.

Brymer, E. (2009). Extreme sports as a facilitator of ecocentricity and positive life changes. *World Leisure Journal* 51 (1) 47–53.

Buckley, R., Gretzel, U., Scott, D., Weaver, D. and Becken, S. (2015) Tourism megatrends. *Tourism Recreation Research* 40 (1), 59–70.

Bull, C. and Weed, M. (1999) Niche markets and small island tourism: The development of sports tourism in Malta. *Managing Leisure* 4 (3), 142–155.

Bunning, R.J. and Gibson, H.J. (2016) The role of travel conditions in cycling tourism: Implications for destination and event management. *Journal of Sport & Tourism* 20 (3–4), 175–194.

Burgan, B. and Mules, T. (1992) Economic impact of sporting events. *Annals of Tourism Research* 19 (4), 700–710.

Burgess, C. (2010) The 'illusion' of homogeneous Japan and national character: Discourse as a tool to transcend the 'myth' vs. 'reality' binary. *The Asia-Pacific Journal | Japan Focus* 8 (9), 1–23.

Burnstyn, V. (1999) *The Rites of Men: Manhood, Politics, and the Culture of Sport.* Toronto: University of Toronto Press.

Burton, R. (1995) *Travel Geography* (2nd edn). London: Pitman Publishing.

Butler, R.W. (1980) The concept of the tourist area lifecycle of evolution: Implications for the management of resources. *Canadian Geographer* 24 (1), 5–12.

Butler, R.W. (1993) Tourism: An evolutionary perspective. In J.G. Nelson, R.W. Butler and G. Wall (eds) *Tourism and Sustainable Development: Monitoring, Planning, Managing* (pp. 27–43). Waterloo, Canada: University of Waterloo – Department of Geography Publication Series No. 37.

Butler, R.W. (1994) Seasonality in tourism: Issues and problems. In A.V. Seaton (ed.) *Tourism: The State of the Art* (pp. 332–339). Chichester: John Wiley and Sons.

Butler, R.W. (1996) The role of tourism in cultural transformation in developing countries. In W. Nuryanti (ed.) *Tourism and Culture: Global Civilization in Change* (pp. 91–101). Yogyakarta: Gadjah Mada University Press.

Butler, R.W. (2001) Seasonality in tourism: Issues and implications. In T. Baum and S. Lundtorp (eds) *Seasonality in Tourism* (pp. 5–23). London: Pergamon.

Butler, R.W. (ed.) (2006) *The Tourist Area Life Cycle: Vol. 1 Applications and Modifications.* Clevedon: Channel View Publications.

Butler, R.W. and Mao, B. (1996) Seasonality in tourism: Problems and measurement. In P.E. Murphy (ed.) *Quality Management in Urban Tourism* (pp. 9–23). Chichester: John Wiley and Sons.

Butler, R.W. and Boyd, S.W. (eds) (2000) *Tourism and National Parks: Issues and Implications*. Chichester: John Wiley and Sons Ltd.

Campelo, A., Aitken, R., Thyne, M. and Gnoth, J. (2013) Sense of place: The importance for destination branding. *Journal of Travel Research* 53 (2), 154–166.

Camy, J., Adamkiewics, E. and Chantelat, P. (1993) Sporting uses of the city: Urban anthropology applied to the sports practices in the agglomeration of Lyon. *International Review for the Sociology of Sport* 29 (2+3), 175–185.

Canadian Sport Tourism Alliance (2017, 2 March) Sport tourism surges past $6.5 billion annually. CSTA Alert, Ottawa.

Canadian Tourism Commission (2002) Sport tourism impact. *Tourism: Canada's Tourism Monthly* 6.

Canadian Tourism Commission & Coopers and Lybrand (1996) *Domestic Tourism Market Research Study*. Ottawa: Canadian Tourism Commission.

Cannas, R. (2012) An overview of tourism seasonality: Key concepts and policies. *Almatourism: Journal of Tourism, Culture and Territorial Development* 3 (5), 40–58.

Cantelon, H. and Gruneau, R.S. (1988) The production of sport for television. In J. Harvey and H. Cantelon (eds) *Not Just a Game: Essays in Canadian Sport Sociology* (pp. 177–194). Ottawa: University of Ottawa Press.

Cantelon, H. and Letters, M. (2000) The making of the IOC environmental policy as the third dimension of the Olympic movement. *International Review for the Sociology of Sport* 35 (3), 294–308.

Carle, A. and Nauright, J. (1999) A man's game? Women playing rugby union in Australia. *Football Studies* 2 (1), 55–73.

Carmichael, B. and Murphy, P.E. (1996) Tourism economic impact of a rotating sports event: The case of the British Columbia Games. *Festival Management and Event Tourism* 4, 127–138.

Carneiro, M.J., Breda, Z. and Cordeiro, C. (2016) Sports tourism development and destination sustainability: The case of the coastal area of the Aveiro region, Portugal. *Journal of Sport & Tourism* 20 (3–4), 305–334.

Carter, J., Dyer, P. and Sharma, B. (2007) Dis-placed voices: Sense of place and place-identity on the Sunshine Coast. *Social and Cultural Geography* 8, 755–773.

Carus, L. (1998) For the strategic analysis and forecasting stage within the strategic tourism management process. *Journal of Sport Tourism* 4 (4), 23–35.

Casey, M.E. (2010) Low cost air travel: Welcome aboard?. *Tourist Studies* 10 (2), 175–191.

Cashmore, E. (1996) *Making Sense of Sports* (2nd edn). London: Routledge.

Castells, M. (1997) *The Rise of the Network Society*. Oxford: Blackwell.

Chadwick, G. (1971) *A Systems View of Planning*. Oxford: Pergamon Press.

Chalip, L. (1992) The construction and use of polysemic structures: Olympic lessons for sport marketers. *Journal of Sport Management* 6, 87–98.

Chalip, L. (2001) Sport tourism: Capitalising on the linkage. In D. Kluka and G. Schilling (eds) *Perspectives: The Business of Sport* (pp. 77–89). Oxford: Meyer and Meyer.

Chalip, L. (2004a) Beyond impact: A general model for host community event leverage. In B. Ritchie and D. Adair (eds) *Sport Tourism: Interrelationships, Impacts and Issues* (pp. 226–252). Clevedon: Channel View Publications.

Chalip, L. (2004b) Olympic teams as market segments. In T.D. Hinch and J.E.S. Higham (eds) *Sport Tourism Development* (pp. 52–54). Clevedon: Channel View Publications.

Chalip, L. (2006) Towards social leverage of sport events. *Journal of Sport & Tourism* 11 (2), 109–127.

Chalip, L. and Costa, C.A. (2005) Sport event tourism and the destination brand: Towards a general theory. *Sport in Society* 8 (2), 218–237.

Chalip, L. and McGuirty, J. (2004) Bundling sport events with the host destination. *Journal of Sport & Tourism* 9 (3), 267–282.

Chalip, L., Green, B.C. and Vander Velden, L. (1998) Sources of interest in travel to the Olympic Games. *Journal of Vacation Marketing* 4 (1), 7–22.

Chalip, L., Green, B.C. and Hill, B. (2003) Effects of sport event media on destination image and intention to visit. *Journal of Sport Management* 17 (3), 214–234.

Chang, P.C. and Singh, K.K. (1990) Risk management for mega-events: The 1988 Olympic Winter Games. *Tourism Management* 11 (1), 45–52.

Chapin, T.S. (2004) Sports facilities as urban redevelopment catalysts: Baltimore's Camden Yards and Cleveland's Gateway. *Journal of the American Planning Association* 70 (2), 193–209.

Chapman, A. and Light, D. (2016) Exploring the tourist destination as a mosaic: The alternative lifecycles of the seaside amusement arcade sector in Britain. *Tourism Management* 52, 254–263.

Chelladurai, P. and Chang, K. (2000) Targets and standards of quality in sport services. *Sport Management Review* 3 (1), 1–22.

Chen, K.C., Groves, D. and Lengfelder, J. (1999) A system model of sport tourism with implications for research. *Visions in Leisure and Business* 18 (1), 34–44.

Chernushenko, D. (1996) Sports tourism goes sustainable: The Lillehammer Experience. *Visions in Leisure and Business* 15 (1), 65–73.

Cho, H., Ramshaw, G. and Norman, W.C. (2014) A conceptual model for nostalgia in the context of sport tourism: Re-classifying the sporting past. *Journal of Sport & Tourism* 19 (2), 145–167.

Chogahara, M. and Yamaguchi, Y. (1998) Resocialization and continuity of involvement in physical activity among elderly Japanese. *International Review for the Sociology of Sport* 33 (3), 277–289.

Christaller, W. (1955) Contributions to the geography of the travel trade. Erkunde Bandix, February 1955. Heft 1. (Regional Science Series).

Christaller, W. (1963/64) Some considerations of tourism location in Europe: The peripheral regions – underdeveloped countries – recreation areas. *Papers, Regional Science Association* 12, 95–105.

Chung, J.Y. (2009) Seasonality in tourism: A review. *E-review of Tourism Research* 7 (5), 82–96.

Clawson, M. and Knetsch, J. (1966) *The Economics of Outdoor Recreation*. Baltimore, MD: Johns Hopkins Press.

Clay, J. (2001, 29 April) Sense of place gives Derby unique allure (p. C1). *Lexington Herald*.

Coakley, J. (2017) *Sports in Society: Issues and Controversies* (12th edn). Boston, MA: McGraw Hill Higher Education.

Coakley, J.L. (1990) *Sport in Society: Issues and Controversies* (4th edn). Missouri, KS: Times Mirror/Mosby College Publishing.

Coalter, F. (1999) Sport and recreation in the United Kingdom: Flow with the flow or buck with the trends? *Managing Leisure* 4 (1), 24–39.

Cohen, E. (1979) A phenomenology of tourist experiences. *Sociology* 13, 179–201.

Cohen, E. (1984) The sociology of tourism: Approaches, issues, and findings. *Annual Review of Sociology* 10 (1), 373–392.

Cohen, E. (1988) Authenticity and the commoditization of tourism. *Annals of Tourism Research* 15, 371–386.

Cohen, E. (1996) A phenomenology of tourist experiences. In Y. Apostolopoulos, S. Leivadi and A. Yiannakis (eds) *The Sociology of Tourism* (pp. 90–111). London: Routledge.

Cohen, E. and Cohen, S. (2012) Authentication: Hot and cool. *Annals of Tourism Research* 39 (3), 1295–1314.

Cole, S. (2007) Beyond authenticity and commodification. *Annals of Tourism Research* 34, 943–960.

Collier, A. (1999) *Principles of Tourism: A New Zealand Perspective* (5th edn). Auckland: Longman.

Collins, A., Jones, C. and Munday, M. (2009) Assessing the environmental impacts of mega sporting events: Two options? *Tourism Management* 30 (6), 828–837.

Collins, M.F. (1991) The economics of sport and sports in the economy: Some international comparisons. In C.P. Cooper (ed.) *Progress in Tourism, Recreation and Hospitality Management* (pp. 184–214). London: Belhaven Press.

Collins, M.F. and Jackson, G. (1999) The economics of sport tourism. In J. Standeven and P. DeKnop (eds) *Sport Tourism* (pp. 170–201). Champaign, IL: Human Kinetics.

Collins, M.F. and Jackson, G. (2001) Evidence for a sports tourism continuum. Paper presented at the Journeys in Leisure: Current and Future Alliances, Luton, UK.

Commons, J. and Page, S. (2001) Managing seasonality in peripheral tourism regions: The case of Northland, New Zealand. In T. Baum and S. Lundtorp (eds) *Seasonality in Tourism* (pp. 153–172). London: Pergamon.

Commonwealth Department of Industry, Science and Resources (2000) Towards a National Sport Tourism Strategy (Draft report). Canberra: Commonwealth Department of Industry, Science and Resources.

Cooper, C., Fletcher, J., Gilbert, D. and Wanhill, S. (1993) *Tourism: Principles and Practice.* Harlow: Longman Group Limited.

Cornelissen, S. (2008) Scripting the nation: Sport, mega-events, foreign policy and state-building in post-apartheid South Africa. *Sport in Society* 11 (4), 481–493.

Cornelissen, S. (2010) The geopolitics of global aspiration: Sport mega-events and emerging powers. *International Journal of the History of Sport* 27 (16–18), 3008–3025.

Cornelissen, S., Bob, U. and Swart, K. (2011) Towards redefining the concept of legacy in relation to sport mega-events: Insights from the 2010 FIFA World Cup. *Development Southern Africa* 28 (3), 307–318.

Cottle, R.L. (1981) Economics of the professional golfers association tour. *Social Science Quarterly* 62 (4), 721–734.

Cowell, R. (1997) Stretching the limits: Environmental compensation, habitat creation and sustainable development. *Transactions of the Institute of British Geographers* 22 (3), 292–306.

Crawford, D.W., Jackson, E.L. and Godbey, G. (1991) A hierarchical model of leisure constraints. *Leisure Sciences* 13, 309–320.

Crawford, S.A.G. (1995) Rugby and the forging of national identity. In J. Nauright (ed.) *Sport, Power and Society in New Zealand: Historical and Contemporary Perspectives* (pp. 5–19). Sydney: University of New South Wales Printery.

Creutzig, F., Jochem, P., Edelenbosch, O.Y., Mattauch, L., van Vuuren, D.P., McCollum, D. and Minx, J. (2015) Transport: A roadblock to climate change mitigation? *Science* 350, 911. doi: 10.1126/science.aac8033.

Crompton, J.L. (1979) Motivations for pleasure vacation. *Annals of Tourism Research* 6 (4), 408–424.

Crompton, J.L. (1995) Economic impact analysis of sports facilities and events: Eleven sources of misapplication. *Journal of Sport Management* 9, 14–35.

Crouch, D. (2000) Places around us: Embodied lay geographies in leisure and tourism. *Leisure Studies* 19 (2), 63–76.

Csikszentmihalyi, M. (1992) *Flow: The Classic Work on How to Achieve Happiness.* London: Rider Paperbacks.

Cuthbertson, B., Heine, M. and Whitson, D. (1997) Producing meaning through movement: An alternative view of sense of place. *Trumpeter* 14 (2), 72–75.

Daniels, M.J. (2007) Central place theory and sport tourism impacts. *Annals of Tourism Research* 34 (2), 332–347.

Dann, G.M.S. (1981) Tourist motivation: An appraisal. *Annals of Tourism Research* 8 (2), 187–219.

Darcy, S. (2003) The politics of disability and access: The Sydney 2000 Games experience. *Disability & Society* 18 (6), 737–757.

Dauncey, H. and Hare, G. (2000) World Cup France '98: Metaphors, meanings and values. *International Review for the Sociology of Sport* 35 (3), 331–347.

Davidson, L. and Stebbins, R. (2011) *Serious Leisure and Nature: Sustainable Consumption in the Outdoors*. New York: Palgrave Macmillan.

Davies, J. and Williment, J. (2008) Sport tourism: Grey sport tourists, all black and red experiences. *Journal of Sport & Tourism* 13 (3), 221–242.

Davis, J. and Thornley, A. (2010) Urban regeneration for the London 2012 Olympics: Issues of land acquisition and legacy. *City, Culture and Society* 1 (2), 89–98.

Dawson, J., Havitz, M. and Scott, D. (2011) Behavioral adaptation of alpine skiers to climate change: Examining activity involvement and place loyalty. *Journal of Travel and Tourism Marketing* 28 (4), 388–404.

Dawson, J., Scott, D. and McBoyle, G. (2009) Climate change analogue analysis of ski tourism in the northeastern USA. *Climate Research* 39 (1), 1–9.

De Knop, P. (1998) Sport tourism: A state of the art. *European Journal for Sport Management* 5 (2), 5–20.

De Melo, V. and Mangan, J.A. (1997) A web of the wealthy: Modern sport in the nineteenth-century culture of Rio de Janeiro. *International Journal of the History of Sport* 14 (1), 168–173.

de Villers, D.J. (2001) Sport and tourism to stimulate development. *Olympic Review* 27 (38), 11–13.

Dear, M. and Flusty, S. (1999) The postmodern urban condition. In M. Featherstone and S. Lash (eds) *Spaces of Culture: City-Nation-World* (pp. 64–85). London: Sage.

Delia, E.B. (2015) The exclusiveness of group identity in celebrations of team success. *Sport Management Review* 18 (3), 396–406.

Deloitte (2014) *All to Play For: Global Money League*. Manchester: Deloitte Sport Business Group.

Delpy, L. (1997) An overview of sport tourism: Building towards a dimensional framework. *Journal of Vacation Marketing* 4 (1), 23–38.

Delpy, L. (1998) Editorial. *Journal of Vacation Marketing* 4 (1), 4–5.

Delpy, L. (2001, 22–23 February) Preparing for the rise in sports tourism. Paper presented at the World Conference on Sport and Tourism, Barcelona, Spain.

Delpy Neirotti, L., Bosetti, H.A. and Teed, K.C. (2001) Motivation to attend the 1996 Summer Olympic Games. *Journal of Travel Research* 39 (3), 327–331.

Denham, D. (2004) Global and local influences on English Rugby League. *Sociology of Sport Journal* 21 (2), 206–182.

Derom, I. and Ramshaw, G. (2016) Leveraging sport heritage to promote tourism destinations: The case of the Tour of Flanders cyclo event. *Journal of Sport & Tourism* 20 (3–4), 263–283.

Devine, A. and Devine, F. (2004) The politics of sports tourism in Northern Ireland. *Journal of Sport Tourism* 9 (2), 171–182.

Dietvorst, A.G.J. (1995) Tourist behaviour and the importance of time-space analysis. In G.J. Ashworth and A.G.J. Dietvorst (eds) *Tourism and Spatial Transformations: Implications for Policy and Planning*. Wallingford: CABI.

Dietvorst, A.G.J. and Ashworth, G.J. (1995) Tourism transformations: An introduction. In G.J. Ashworth and A.G.J. Dietvorst (eds) *Tourism and Spatial Transformations: Implications for Policy and Planning* (pp. 1–13). Wallingford: CABI.

Dixon, K. (2014) The role of lateral surveillance in the construction of authentic football fandom practice. *Surveillance & Society* 11 (4), 424–438.

Dodds, R. and Graci, S. (2009) Canada's tourism industry—Mitigating the effects of climate change: A lot of concern but little action. *Tourism and Hospitality Planning and Development* 6 (1), 39–51.

Doering, A. (2018) Mobilising stoke: A genealogy of surf tourism development in Miyazaki, Japan. *Tourism Planning & Development* 15 (1), 68–81.

Dolnicar, S. (2002) A review of data-driven market segmentation in tourism. *Journal of Travel & Tourism Marketing* 12 (1), 1–22.

Donnelly, P. and Young, K.M. (1985) Reproductions and transformation of cultural forms in sport: A contextual analysis of rugby. *International Journal of the History of Sport* 20, 19–37.

Donnelly, P. and Young, K.M. (1988) The construction and confirmation of identity in sport subcultures. *Sociology of Sport Journal* 5, 223–240.

Doxey, G. (1975, 8–11 September) Visitor–resident interaction in tourist destinations: Inferences from empirical research in Barbados, West Indies and Niagara-on-the-Lake, Ontario. Paper presented at the Symposium on the Planning and Development of the Tourist Industry in the ECC Region, Dubrovnik, Yugoslavia.

Duda, J.L. and Nicholls, J.G. (1992) Dimensions of achievement motivation in schoolwork and sport. *Journal of Educational Psychology*, 84 (3), 290.

Duncan, M. and Brummett, B. (1989) Types and sources of spectating pleasure in televised sports. *Sociology of Sport Journal* 6 (3), 195–211.

Dunn Ross, E.L. and Iso-Ahola, S.E. (1991) Sightseeing tourists' motivation and satisfaction. *Annals of Tourism Research* 18, 226–237.

Dunning, E. (1994) Sport in space and time: 'Civilizing processes', trajectories of state-formation and the development of modern sport. *International Review for the Sociology of Sport* 29 (4), 331–348.

Dunning, E. (1999) *Sport Matters: Sociological Studies of Sport, Violence and Civilisation.* London: Routledge.

Dutch Ministry of Economic Affairs (1991) *Improving Seasonal Spread of Tourism.* Rotterdam: Markant-Adviesbureau, Dutch Ministry of Economic Affairs.

Dwyer, L. (2014) Transnational corporations and the globalization of tourism. In A.A. Lew, C.M. Hall and A.M. Williams (eds) *The Wiley Blackwell Companion to Tourism* (pp. 197–209). Chichester: John Wiley & Sons.

Dwyer, L., Edwards, D., Mistilis, N., Roman, C., Scott. N. and Cooper, C. (2008) Megatrends underpinning tourism to 2020: Analysis of key drivers for change. CRC for Sustainable Tourism Pty Ltd. See http://crctourism.com.au/WMS/Upload/Resources/bookshop/80046%20Dwyer_TourismTrends2020%20WEB.pdf (accessed 23 April 2010).

Eastman, S.T. and Riggs, K.E. (1994) Televised sports and ritual: Fan experiences. *Sociology of Sport Journal* 11 (3), 249–274.

Echtner, C.M. and Ritchie, J.B.R. (1993) The measurement of destination image : An empirical assessment. *Journal of Travel Research* Spring, 3–13.

Eden, S. and Barratt, P. (2010) Outdoors versus indoors? Angling ponds, climbing walls and changing expectations of environmental leisure. *Area* 42 (4), 487–493.

Edensor, T. and Richards, S. (2007) Snowboarders vs skiers: Contested choreographies of the slopes. *Leisure Studies* 26 (1), 97–114.

Edensor, T. and Millington, S. (2008) 'This is our city': Branding football and local embeddedness. *Global Networks: A Journal of Transnational Affairs* 8 (2), 172–193.

Edgell, D.L. (1990) *International Tourism Policy.* New York: Van Nostrand Reinhold.

Eid, M. and Diener, E. (2001) Norms for experiencing emotions in different cultures: Inter- and intranational differences. *Journal of Personality and Social Psychology* 81 (5), 869–884.

Elias, N. and Dunning, E. (2008) *Quest for Excitement: Sport and Leisure in the Civilising Process* (rev. edn). Dublin: University College Dublin Press.

Elling, A. and Janssens, J. (2009) Sexuality as a structural principle in sport participation: Negotiating sports spaces. *International Review for the Sociology of Sport* 44 (1), 71–86.

Emery, P.R. (1998, 2–4 July) Bidding to host a major sports event: Strategic investment or complete lottery. Paper presented at the Sport in the City, Sheffield, UK.

Environment Agency (2008) Assessing Optimum Irrigation Water Use: Additional Agricultural and Non-Agricultural Sectors (Report No. SC040008/SR1). Bristol: Environment Agency.

Ernst and Young (1996) Economic Impact Analysis America's Cup Auckland 2000. Unpublished report prepared for the Auckland Regional Services Trust, Auckland.

Esfahani, N., Goudarzi, M. and Assadi, H. (2009) The analysis of the factors affecting the development of Iran sport tourism and the presentation of a strategic model. *World Journal of Sport Sciences* 2, 136–144.

Evans, D. and Norcliffe, G. (2016) Local identities in a global game: The social production of football space in Liverpool. *Journal of Sport & Tourism* 20 (3–4), 217–232.

Ewert, A. and Shultis, J. (1999) Technology and backcountry recreation: Boon to recreation or bust for management. *Journal of Physical Education, Recreation and Dance* 70 (8), 22–31.

Fairley, S. (2003) In search of relived social experience: Group-based nostalgia sport tourism. *Journal of Sport Management* 17 (3), 284–304.

Fairley, S. and Gammon, S. (2005) Something lived, something learned: Nostalgia's expanding role in sport tourism. *Sport in Society* 8 (2), 182–197.

Fairley, S. and Tyler, B.D. (2009) Cultural learning through a sport tourism experience: The role of the group. *Journal of Sport Tourism* 14 (4), 273–292.

Falcous, M. (2017) Why we ride: Road cyclists, meaning, and lifestyles. *Journal of Sport and Social Issues* 41 (3), 239–255.

Falcous, M. and Newman, J.I. (2016) Sporting mythscapes, neoliberal histories, and post-colonial amnesia in *Aotearoa*/New Zealand. *International Review for the Sociology of Sport* 51 (1), 61–77.

Falla, J. (2000) *Home Ice: Reflections on Backyard Rinks and Frozen Ponds*. Toronto: McClelland & Stewart.

Fang, Y. and Yin, J. (2015) National assessment of climate resources for tourism seasonality in China using the tourism climate index. *Atmosphere* 6, 183–194.

Faulkner, B., Tideswell, C. and Weston, A.M. (1998) Leveraging tourism benefits from the Sydney 2000 Olympics. Paper presented at the Sport Management Association of Australia and New Zealand, Gold Coast, Australia, 26–28 November.

FC Barcelona (2016) FC Barcelona Museum preparing for 30 millionth visitor. See https://www.fcbarcelona.com/club/news/2016-2017/fc-barcelona-museum-preparing-for-30-millionth-visitor (accessed 14 July 2017).

Fejgin, N. (1994) Participation in high school competitive sports: A subversion of school mission or contribution to academic goals? *Sociology of Sport Journal* 11 (3), 211–230.

Finney, B.R. and Houston, J.D. (1996) *Surfing: A History of the Ancient Hawaiian Sport*. San Francisco, CA: Pomegranate.

Firat, A. (1995) Consumer culture or culture consumed. In J. Costa and G. Bamossy (eds) *Marketing in a Multicultural World* (pp. 105–123). London/Thousand Oaks, CA: Sage.

Flagestad, A. and Hope, C.A. (2001) Strategic success in winter sports destinations: A sustainable value creation perspective. *Tourism Management* 22 (5), 445–461.

Flognfeldt, T. (2001) Long-term positive adjustments to seasonality: Consequences of summer tourism in the Jotunheimen area, Norway. In T. Baum and S. Lundtorp (eds) *Seasonality in Tourism* (pp. 109–118). Oxford: Pergamon.

Fougere, G. (1989) Sport, culture and identity: The case of rugby football. In D. Novitz and B. Willmott (eds) *Cultural Identity in New Zealand* (pp. 110–122). Wellington: GP Books.

Fourie, J. and Santana-Gallego, M. (2011) The impact of mega-sport events on tourist arrivals. *Tourism Management* 32 (6), 1364–1370.

Francis, S. and Murphy, P.E. (2005) Sport tourism destinations: The active sport tourist perspective. In J.E.S. Higham (ed.) *Sport Tourism Destinations: Issues, Opportunities and Analysis* (pp. 73–92). Oxford: Elsevier.

Frechtling, D.C. (1996) *Practical Tourism Forecasting*. Oxford: Butterworth-Heinemann.

Freidmann, J. (1986) The world city hypothesis. *Development and Change* 17, 69–84.

Frosdick, S. and Marsh, P. (2013) *Football Hooliganism*. London: Routledge.

Funk, D.C. and Bruun, T.J. (2007) The role of socio-psychological and culture-education motives in marketing international sport tourism: A cross-cultural perspective. *Tourism Management* 28 (3), 806–819.

Fyall, A. and Jago, L. (eds) (2009) Sustainability in sport and tourism. *Journal of Sport & Touirsm* 14 (2–3), 77–81.

Gaboriau, P. (2003) The Tour de France and cycling's Belle Epoque. *International Journal of the History of Sport* 20 (2), 57–78.

Gallarza, M.G., Saura, I. and Garcia, H. (2001) Destination image: Towards a conceptual framework. *Annals of Tourism Research* 29 (1), 56–78.

Gammon, S. (2002) Fantasy, nostalgia and the pursuit of what never was. In S. Gammon and J. Kurtzman (eds) *Sport Tourism: Principles and Practice* (pp. 61–72). Eastbourne: Leisure Studies Association.

Gammon, S. (2015) Sport tourism finding its place? In S. Gammon and S. Elkington (eds) *Landscapes of Leisure* (pp. 110–122). London: Palgrave Macmillan.

Gammon, S. and Fear, V. (2005) Stadia tours and the power of backstage. *Journal of Sport Tourism* 10 (4), 243–252.

Gammon, S. and Kurtzman, J. (eds) (2002) *Sport Tourism: Principles and Practice*. Eastbourne: Leisure Studies Association.

Gammon, S. and Ramshaw, G. (2007) *Heritage, Sport and Tourism: Sporting Pasts–Tourist Futures*. New York: Routledge.

Gammon, S. and Ramshaw, G. (2013) *Heritage, Sport and Tourism: Sporting Pasts–Tourist Futures*. London: Routledge.

Gammon, S. and Robinson, T. (1997) Sport and tourism: A conceptual framework. *Journal of Sport Tourism* 4 (3), 8–24.

Gammon, S. and Robinson, T. (1999) The development and design of the sport tourism curricular with particular references to the BA (hons) sport tourism degree at the University of Luton. *Journal of Sport Tourism* 5 (2), 17–26.

Gammon, S. and Robinson, T. (2003) Sport and tourism: A conceptual framework. *Journal of Sport & Tourism* 8, 21–26.

Gammon, S., Ramshaw, G. and Waterton, E. (2013) Examining the Olympics: Heritage, identity and performance. *International Journal of Heritage Studies* 19 (2), 119–124.

Gammon, S., Ramshaw, G. and Wright, R. (2017) Theory in sport tourism: Some critical reflections. *Journal of Sport & Tourism* 21 (2), 69–74.

Gantz, W. and Wenner, L.A. (1995) Fanship and the television sports viewing experience. *Sociology of Sport Journal* 12 (1), 56–74.

Garau-Vadell, J.B. and de Borja-Sole, L. (2008) Golf in mass tourism destinations facing seasonality: A longitudinal study. *Tourism Review* 63 (2), 16–24.

García, B. (2010) The concept of the Olympic Cultural Programme: Origin, evolution and projection. Centre d'Estudis Olímpics (CEO-UAB), Barcelona. International Chair in Olympism (IOC–UAB). See http://ceo.uab.cat.

Garmise, M. (1987) *Proceedings of the International Seminar and Workshop on Outdoor Education, Recreation and Sport Tourism*. Natanya: Emmanuel Gill Publishing.

Garrod, B. (2009) Understanding the relationship between tourism destination imagery and tourist photography. *Journal of Travel Research* 47 (3), 346–358.

Gee, S. (2014) Sport and alcohol consumption as a neoteric moral panic in New Zealand: Context, voices and control. *Journal of Policy Research in Tourism, Leisure and Events* 6 (2), 153–171.

Getz, D. (1991) *Festivals, Special Events and Tourism.* New York: Van Nostrand Reinhold.

Getz, D. (1997) *Event Management and Event Tourism.* New York: Cognizant Communications Corporation.

Getz, D. (1998) Trends, strategies, and issues in sport-event tourism. *Sport Marketing Quarterly* 7 (2), 8–13.

Getz, D. (2008) Event tourism: Definition, evolution, and research. *Tourism Management* 29 (3), 403–428.

Getz, D. and Cheyne, J. (1997) Special event motivations and behaviour. In C. Ryan (ed.) *The Tourist Experience: A New Introduction* (pp. 136–154). London: Cassell.

Getz, D. and McConnell, A. (2011) Serious sport tourism and event travel careers. *Journal of Sport Management* 25 (4), 326–338.

Getz, D. and McConnell, A. (2014) Comparing trail runners and mountain bikers: Motivation, involvement, portfolios, and event-tourist careers. *Journal of Convention & Event Tourism* 15 (1), 69–100.

Getz, D. and Page, S.J. (2016) *Event Studies: Theory, Research and Policy for Planned Events* (3rd edn). London: Routledge.

Ghaderi, Z., Khoshkam, M. and Henderson, J.C. (2014 From snow skiing to grass skiing: Implications of climate change for the ski industry in Dizin, Iran. *Anatolia* 25 (1), 96–107.

Gibson, H.J. (1998a) Sport tourism: A critical analysis of research. *Sport Management Review* 1 (1), 45–76.

Gibson, H.J. (1998b) Active sport tourism: Who participates? *Leisure Studies* 17, 155–170.

Gibson, H.J. (1998c) The wide world of sport tourism. *Parks and Recreation* 33 (9), 108–114.

Gibson, H.J. (2002) Sport tourism at a crossroad? Considerations for the future. In S. Gammon and J. Kurtzman (eds) *Sport Tourism: Principles and Practice* (Vol. 76; pp. 111–128). Eastbourne: Leisure Studies Association.

Gibson, H.J. (2005) Understanding sport tourism experiences. In J.E.S. Higham (ed.) *Sport Tourism Destinations: Issues, Opportunities and Analysis* (pp. 57–72). Oxford: Elseiver Butterworth Heinemann.

Gibson, H.J. (ed.) (2006) *Sport Tourism: Concepts and Theories.* London: Routledge.

Gibson, H.J., Attle, S.P. and Yiannakis, A. (1998) Segmenting the active sport tourist market: A life-span perspective. *Journal of Vacation Marketing* 4 (1), 52–64.

Gibson, H.J., Willming, C. and Holdnak, A. (2002) Small-scale event sport tourism: College sport as a tourist attraction. In S. Gammon and J. Kurtzman (eds) *Sport Tourism: Principles and Practice* (pp. 3–18). Eastbourne: Leisure Studies Association.

Gibson, H.J., Kaplanidou, K. and Kang, S.J. (2012) Small-scale event sport tourism: A case study in sustainable tourism. *Sport Management Review* 15 (2), 160–170.

Gibson, O. (2012) Premier League Lands £3bn TV Rights Bonanza from Sky and BT. *The Guardian.* See http://www.theguardian.com/media/2012/jun/13/premier-league-tv-rights-3-billion-sky-bt (accessed 24 June 2017).

Gilbert, D. and Hudson, S. (2000) Tourism demand constraints: A skiing participation. *Annals of Tourism Research* 27 (4), 906–925.

Gilchrist, P. and Wheaton, B. (2011) Lifestyle sport, public policy and youth engagement: Examining the emergence of parkour. *International Journal of Sport Policy and Politics* 3 (1), 109–131.

Gilchrist, P. and Wheaton, B. (2016) Lifestyle and adventure sports among youth. In K. Green and A. Smith (eds) *Routledge Handbook on Youth Sport* (Chapter 18). New York: Routledge.

Gillespie, D., Leffler, A. and Lerner, E. (2002) If it weren't for my hobby, I'd have a life: Dog sports, serious leisure, and boundary negotiations. *Leisure Studies* 21, 285–304.

Gillett, P. and Kelly, S. (2006) 'Non-local' Masters Games participants: An investigation of competitive active sport tourist motives. *Journal of Sport Tourism* 11 (3–4), 239–257.

Gilmore, J.H. and Pine, B.J. (2007) *Authenticity: What Consumers Really Want*. Boston, MA: Harvard Business School Press.

Giulianotti, R. (1991) Scotland's tartan army in Italy: The case for the carnivalesques. *Sociological Review* 39 (3), 503–527.

Giulianotti, R. (1995a) Football and the politics of carnival: An ethnographic study of Scottish fans in Sweden. *International Review for the Sociology of Sport* 30 (2), 191–223.

Giulianotti, R. (1995b) Participant observation and research into football hooliganism: Reflections on the problems of entree and everyday risks. *Sociology of Sport Journal* 12 (1), 1–20.

Giulianotti, R. (1996) Back to the future: An ethnography of Ireland's football fans at the 1994 World Cup finals in the USA. *International Review for the Sociology of Sport* 31 (3), 323–347.

Giulianotti, R. (2016) *Sport: A Critical Sociology* (2nd edn). Oxford: Polity Press.

Glyptis, S.A. (1982) *Sport and Tourism in Western Europe*. London: British Travel Education Trust.

Glyptis, S.A. (1989) Leisure and patterns of time use. Paper presented at the Leisure Studies Association Annual Conference, Bournemouth, England, 24–26 April 1987.

Glyptis, S.A. (1991) Sport and tourism. In C.P. Cooper (ed.) *Progress in Tourism, Recreation and Hospitality Management* (pp. 165–187). London: Belhaven Press.

Go, F.M. (2004) Tourism in the context of globalization. In S. Williams (ed.) *Tourism: Critical Concepts in the Social Sciences* (pp. 49–80). London: Routledge.

Godbey, G. and Graefe, A. (1991) Repeat tourism, play and monetary spending. *Annals of Tourism Research* 18 (2), 213–225.

Gold, J.R. and Gold, M.M. (eds) (2016) *Olympic Cities: City Agendas, Planning, and the World's Games, 1896–2020*. London: Routledge.

Goldman, R. and Papson, S. (1998) *Nike Culture: The Sign of the Swoosh*. London: Sage.

Golf Canada (2014) Sustainability pilot. See http://www.golfcanadafoundation.com/partners/sustainability-pilot/ (accessed 18 June 2017).

Gomez Martin, M.B. (2005) Weather, climate and tourism a geographic perspective. *Annals of Tourism Research* 32, 571–591.

Goulding, C. (1999) Heritage, nostalgia, and the 'grey' consumer. *Journal of Marketing Practice: Applied Marketing Science* 5 (6), 177–199.

Grabowski, P. (1999) Tourism and sport: Parallel tracks for developing tourism in Brunei? *Tourism Recreation Research* 24 (2), 95–98.

Graburn, N.H.H. (1989) Tourism: The sacred journey. In V.L. Smith (ed.) *Hosts and Guests: The Anthropology of Tourism* (2nd edn). Philadelphia, PA: University of Pennsylvania Press.

Graczyk, W. (2014) Foreign fans revel in ballpark fun. *The Japan Times*. http://www.japantimes.co.jp/sports/2014/05/17/baseball/foreign-fans-revel-ballpark-fun/#.WOSdAPnyiiM (accessed 1 April 2017).

Graefe, A.R., Vaske, J.J. and Kuss, F.R. (1984) Social carrying capacity: An integration and synthesis of twenty years of research. *Leisure Sciences* 6 (4), 395–431.

Grainger, A. and Jackson, S. (1999) Resiting the swoosh in the land of the long white cloud. *Peace Review* 11 (4), 511–516.

Grainger, A. and Jackson, S. (2000) Sports marketing and the challenges of globalization: A case study of cultural resistance in New Zealand. *International Journal of Sports Marketing & Sponsorship* 2 (2), 35–49.

Gratton, C., Dobson, N. and Shibli, S. (2000) The economic importance of major sports events: A case-study of six events. *Managing Leisure* 5 (1), 17–28.

Gratton, C. and Preuss, H. (2008) Maximizing Olympic impacts by building up legacies. *International Journal of the History of Sport* 25 (14), 1922–1938.

Gratton, C., Shibli, S. and Coleman, R. (2005) Sport and economic regeneration in cities. *Urban Studies* 42 (5–6), 985–999.

Gratton, C., Shibli, S. and Coleman, R. (2006) The economic impact of major sports events: A review of ten events in the UK. *The Sociological Review* 54 (s2), 41–58.

Gratton, C. and Taylor, P. (2000) *Economics of Sport and Recreation*. London: E&FN Spon.

Green, B.C. (2001) Leveraging subculture and identity to promote sport events. *Sport Management Review* 4 (1), 1–19.

Green, B.C. and Chalip, L. (1998) Sport tourism as the celebration of subculture. *Annals of Tourism Research* 25 (2), 275–291.

Green, P. (1992) *Alexander of Macedon, 356–323 BC: A Historical Biography*. Berkeley, CA: University of California Press.

Greenwood, D.J. (1989) Culture by the pound: An anthropological perspective on tourism as cultural commodification. In V.L. Smith (ed.) *Hosts and Guests: The Anthropology of Tourism* (pp. 17–31). Philadelphia, PA: University of Pennsylvania Press.

Groff, D., Funderburk, J., McComb, A. and Connolly, S. (2000) Ninety minutes into the game. *Parks and Recreation* 35 (8), 70–78.

Grundlingh, A. (1994) Playing for power? Rugby, Afrikaner nationalism and masculinity in South Africa, c.1900–70. *International Journal of the History of Sport* 11 (3), 408–430.

Gu, H. and Ryan, C. (2008) Place attachment, identity and community impacts of tourism: The case of a Beijing hutong. *Tourism Management* 29, 637–647.

Gunn, C. (1988) *Vacationscape: Designing Tourist Regions* (2nd edn). New York: Van Nostrand Reinhold.

Gustafson, S. (2013) Displacement and the racial state in Olympic Atlanta. *Southeastern Geographer* 53 (2), 198–213.

Guttmann, A. (1992) *The Olympics: A History of the Modern Games*. Urbana, IL: University of Illinois Press.

Hackworth, J. (2008) The durability of roll-out neoliberalism under centre-left governance: The case of Ontario's social housing sector. *Studies in Political Economy* 21, 7–26.

Hagen, S. and Boyes, M. (2016) Affective ride experiences on mountain bike terrain. *Journal of Outdoor Recreation and Tourism* 15, 89–98.

Halberstam, D. (1999) *Playing for Keeps: Michael Jordan and the World He Made*. New York: Random House.

Hall, C.M. (1992a) *Hallmark Tourist Events: Impacts, Management and Planning*. London: Belhaven Press.

Hall, C.M. (1992b) Review: Adventure, sport and health tourism. In B. Weiler and C.M. Hall (eds) *Special Interest Tourism* (pp. 186–210). London: Belhaven Press.

Hall, C.M. (1993) The politics of leisure: An analysis of spectacles and mega-events. In A.J. Veal, P. Johnson and G. Cushman (eds) *Leisure and Tourism: Social and Environmental Changes* (pp. 620–629). Sydney: World Leisure and Recreation Association.

Hall, C.M. (1995) *Introduction to Tourism in Australia: Impacts, Planning and Development* (2nd edn). South Melbourne: Addison Wesley Longman Australia.

Hall, C.M. (1998) Imaging, tourism and sports event fever: The Sydney Olympics and the need for a social charter for mega-events. In C. Gratton and I.P. Henry (eds) *Sport in the City: The Role of Sport in Economic and Social Regeneration* (pp. 166–183). London: Routledge.

Hall, C.M. (1999) Rethinking collaboration and partnership: A public policy perspective. *Journal of Sustainable Tourism* 7 (3–4), 274–289.

Hall, C.M. (2000a) *Tourism Planning: Policies, Processes and Relationships*. Harlow: Prentice-Hall.

Hall, C.M. (2000b) The future of tourism: A personal speculation. *Tourism Recreation Research* 25 (1), 85–95.

Hall, C.M. (2004) Sport tourism and urban regeneration. In B. Ritchie and D. Adair (eds) *Sport Tourism: Interrelationships, Impacts and Issues* (pp. 192–205). Clevedon: Channel View Publications.

Hall, C.M. (2008) *Tourism Planning: Policies, Processes and Relationships*. Harlow: Pearson Education.

Hall, C.M. and Weiler, B. (eds) (1992) *Special Interest Tourism*. London: Belhaven Press.

Hall, C.M. and Lew, A.A. (eds) (1998) *Sustainable Tourism: A Geographical Perspective*. Harlow: Addison Wesley Longman Ltd.

Hall, C.M. and Higham, J.E.S. (eds) (2005) *Tourism, Recreation and Climate Change: International Perspectives*. Clevedon: Channel View Publications.

Hall, C.M. and Page, S.J. (2014) *The Geography of Tourism and Recreation: Environment, Place and Space*. London: Routledge.

Hall, C.M., Jenkins, J. and Kearsley, G.W. (eds) (1997) *Tourism Planning and Policy in Australia and New Zealand*. Sydney: Irwin Publishers.

Hallmann, K., Zehrer, A. and Müller, S. (2015) Perceived destination image: An image model for a winter sports destination and its effect on intention to revisit. *Journal of Travel Research* 54 (1), 94–106.

Halpenny, E.A., Kulczycki, C. and Moghimehfar, F. (2016) Factors effecting destination and event loyalty: Examining the sustainability of a recurrent small-scale running event at Banff National Park. *Journal of Sport & Tourism* 20 (3–4), 233–262.

Hamilton, L.C., Brown, C. and Keim, B.D. (2007) Ski areas, weather and climate: Time series models for New England case studies. *International Journal of Climatology* 27, 2113–2124.

Hammitt, W.E. (1980) Outdoor recreation: Is it a multi-phase experience? *Journal of Leisure Research* 12 (2), 107–115.

Hanna, S. and Rowley, J. (2008) An analysis of terminology use in place branding. *Place Branding and Public Diplomacy* 4 (1), 61–75.

Harada, M. (2016) *Supotsu toshi senryaku* [*Strategic Planning for the Development of the Sports City*]. Kyoto: Gakugei Publisher (in Japanese).

Harahousou, Y. (1999) Elderly people, leisure and physical recreation in Greece. *World Leisure and Recreation* 41 (3), 20–24.

Harrison, S.J., Winterbottom, W.J. and Shepard, C. (1999) The potential effects of climate change on the Scottish tourism industry. *Tourism Management* 20 (2), 25–33.

Harrison-Hill, T. and Chalip, L. (2005) Marketing sport tourism: Creating synergy between sport and destination. *Sport in Society* 8 (2), 302–320.

Hartman, R. (1986) Tourism, seasonality and social change. *Leisure Studies* 5 (1), 25–33.

Harvey, D. (1989) *The Condition of Postmodernity*. Oxford: Blackwell Publishers Inc.

Harvey, D. (2007) Neoliberalism as creative destruction. *The Annals of the American Academy of Political and Social Science* 610 (1), 21–44.

Harvey, J. and Houle, F. (1994) Sport, world economy, global culture, and new social movements. *Sociology of Sport Journal* 11 (4), 337–355.

Harvey, J., Rail, G. and Thibault, I. (1996) Globalization and sport: Sketching a theoretical model for empirical analyses. *Journal of Sport and Social Issues* 23 (3), 258–277.

Harvey, J., Horne, J., Safai, P., Darnell, S. and Courchesne-O'Neill, S. (2013) *Sport and Social Movements: From the Local to the Global*. London: Bloomsbury.

Hawkins, D.E. and Mann, S. (2007) The World Bank's role in tourism development. *Annals of Tourism Research* 34, 348–363.

Hawkins, P. (1999) Sports Tourism in the Peak National Park. In M. Scarrott (ed.) *Exploring Sports Tourism: Proceedings of a SPRIG Seminar Held at the University of Sheffield on 15 April 1999* (pp. 38–45). Sheffield: Sheffield Hallam University.

Heath, E.T. and Kruger, E.A. (2015) Spectators' contribution to the environmental dimension of sustainable event sports tourism. Unpublished PhD thesis, University of Pretoria.

Hede, A.M. and Kellett, P. (2010) Why develop Melbourne Park? In T.D. Hinch and J.E.S. Higham (eds) *Sport Tourism Development* (2nd edn). Bristol: Channel View Publications.

Hein, L., Metzger, M. and Moren, A. (2009) Potential impacts of climate change on tourism: A case study for Spain. *Current Opinion in Environmental Sustainability* 1, 170–178.

Heino, R. (2000) What is so punk about snowboarding? *Journal of Sport and Social Issues* 24 (1), 176–191.

Heitzman, J. (1999) Sports and conflict in urban planning. The Indian national games in Bangalore. *Journal of Sport and Social Issues* 23 (1), 5–23.

Henderson, J.C., Foo, K., Lim, H. and Yip, S. (2010) Sports events and tourism: The Singapore formula one grand prix. *International Journal of Event and Festival Management* 1 (1), 60–73.

Hendrikx, J. and Hreinsson, E.Ö. (2012) The potential impact of climate change on seasonal snow in New Zealand: Part II—industry vulnerability and future snowmaking potential. *Theoretical & Applied Climatology* 110, 607–618.

Heslop, L.A., Nadeau, J., O'Reilly, N. and Armenakyan, A. (2013) Mega-event and country co-branding: Image shifts, transfers and reputational impacts. *Corporate Reputation Review* 16 (1), 7–33.

Higgins, M. (2016, 6 March) Snowboarding, Once a High-Flying Sport, Crashes to Earth. *The New York Times*. See https://www.nytimes.com/2016/03/07/sports/snowboarding-once-a-high-flying-sport-crashes-to-earth.html?_r=0 (accessed 24 June 2017).

Higham, J.E.S. (1996) The Bledisloe Cup: Quantifying the direct economic benefits of event tourism, with ramifications for a city in economic transition. *Festival Management and Event Tourism* 4, 107–116.

Higham, J.E.S. (1999) Sport as an avenue of tourism development: An analysis of the positive and negative impacts of sport tourism. *Current Issues in Tourism* 2 (1), 82–90.

Higham, J.E.S. (2005) *Sport Tourism Destinations: Issues, Opportunities and Analysis.* Oxford: Elsevier Butterworth Heinemann.

Higham, J.E.S., Cohen, S.A., Cavaliere, C.T., Reis, A.C. and Finkler, W. (2016) Climate change, tourist air travel and radical emissions reduction. *Journal of Cleaner Production* 111, 336–347.

Higham, J.E.S. and Hinch, T.D. (1998) The transition to professional rugby union in New Zealand: An analysis of the temporal dimensions of tourism within the Otago Highlanders franchise. Paper presented at the Proceedings of the New Zealand Tourism and Hospitality Research Conference (Part I), Akaroa, New Zealand, 1–4 December.

Higham, J.E.S. and Hinch, T.D. (2000) Sport tourism and the transition to professional Rugby Union in New Zealand: The spatial dimension of tourism associated with the Otago Highlanders, Southern New Zealand. In P.L.M. Robinson, N. Evans, R. Sharpley and J. Swarbrooke (eds) *Reflections on International Tourism: Motivations, Behaviour and Tourists Types* (Vol. 4; pp. 145–158). Sunderland: Business Education Publishers Ltd.

Higham, J.E.S. and Hinch, T.D. (2002a) Sport, tourism and seasons: The challenges and potential of overcoming seasonality in the sport and tourism sectors. *Tourism Management* 23, 175–185.

Higham, J.E.S. and Hinch, T.D. (2002b) Sport and tourism development: Avenues of tourism development associated with a regional sport franchise at an urban tourism destination. In S. Gammon and J. Kurtzman (eds) *Sport Tourism: Principles and Practice* (pp. 19–34). Eastbourne: Leisure Studies Association.

Higham, J.E.S. and Hinch, T.D. (2003) Sport, space and time: Effects of the Otago Highlanders franchise on tourism. *Journal of Sports Management* 17 (3), 235–257.

Higham, J.E.S. and Hinch, T.D. (2006) Sport and tourism research: A geographic approach. *Sport & Tourism: A Multidisciplinary Journal* 11 (1), 31–49.

Higham, J.E.S. and Hinch, T.D. (2010) *Sport and Tourism: Globalisation, Mobility and Identity.* Oxford: Butterworth Heinemann.

Hill, C.R. (1992) *Olympic Politics.* Manchester: Manchester University Press.

Hill, J.S. and Vincent, J. (2006) Globalisation and sports branding: The case of Manchester United. *International Journal of Sports Marketing and Sponsorship* 7 (3), 61–78.

Hill, J. and McLean, D.C. (1999) Introduction: Possible, probable, or preferable future? *Journal of Physical Education, Recreation & Dance* 70 (9), 15–17.

Hiller, H.H. (1998) Assessing the impacts of mega-events: A linkage model. *Current Issues in Tourism* 1 (1), 47–57.

Hinch, T.D. (2006) Canadian sport and culture in the tourism marketplace. *Tourism Geographies* 8 (1), 15–30.

Hinch, T.D. (2013) Ultra-marathons and tourism development: The case of the Canadian death race in Grande Cache, Alberta. In B. Garrod and A. Fyall (eds) *Contemporary Cases in Sport* (pp. 22–40). Oxford: Goodfellow Publishers.

Hinch, T., Hickey, G. and Jackson, E.L. (2001) Seasonal visitation at Fort Edmonton Park: An empirical analysis using a leisure constraints framework. In T. Baum and S. Lundtorp (eds) *Seasonality in Tourism* (pp. 173–186). London: Pergamon.

Hinch, T.D. and Higham, J.E.S. (2001) Sport tourism: A framework for research. *The International Journal of Tourism Research* 3 (1), 45–58.

Hinch, T.D. and Higham, J.E.S (2005) Sport, tourism and authenticity. *European Sport Management Quarterly: Special Issue Sports Tourism Theory and Method* 5 (3), 243–256.

Hinch, T.D. and Higham, J.E.S. (2004) *Sport Tourism Development.* Clevedon: Channel View Publications.

Hinch, T.D., Higham, J.E.S. and Doering, A. (2018) Sport, tourism and identity: Japan, rugby union and the transcultural maul. In C. Acton and D. Hassan (eds) *Sport and Contested Identities: Contemporary Issues and Debates* (pp. 191–206). London and New York: Routledge.

Hinch, T.D., Higham, J.E.S. and Moyle, B.D. (2016) Sport tourism and sustainable destinations: Foundations and pathways. *Journal of Sport & Tourism* 20 (3–4), 163–173.

Hinch, T., Higham, J. and Sant, S.L. (2014) Taking stock of sport tourism research. In A. Lew, C.M. Hall and A.M. Williams (eds) *The Wiley Blackwell Companion to Tourism* (pp. 414–424). Chichester: John Wiley.

Hinch, T. and Holt, N.L. (2017) Sustaining places and participatory sport tourism events. *Journal of Sustainable Tourism* 25 (8), 1084–1099.

Hinch, T. and Ito, E. (2018) Sustainable sport tourism in Japan. *Tourism Planning & Development* 15 (1), 96–101.

Hinch, T.D. and Jackson, E.L. (2000) Leisure constraints research: Its value as a framework for understanding tourism seasonality. *Current Issues in Tourism* 3 (2), 87–106.

Hinch, T. and Kono, S. (2017) Ultramarathon runners' perception of place: A photo-based analysis. *Journal of Sport & Tourism.* doi: http://dx.doi.org/10.1080/14775085.2017.1371065.

Hjalager, A. (2007) Stages in the economic globalisation of tourism. *Annals of Tourism Research* 34, 437–457.

Hodeck, A. and Hovemann, G. (2016) Motivation of active sport tourists in a German highland destination: A cross-seasonal comparison. *Journal of Sport & Tourism* 20 (3–4), 335–348.

Hodge, K. and Hermansson, G. (2007) Psychological preparation of athletes for the Olympic context: The New Zealand summer and winter Olympic teams. *Athletic Insight* 9 (4), 1–14.

Hodge, K., Lonsdale, C. and Ng, J.Y. (2008) Burnout in elite rugby: Relationships with basic psychological needs fulfilment. *Journal of Sports Sciences* 26 (8), 835–844.

Hodge, K., Lonsdale, C. and Oliver, A. (2010) The elite athlete as a 'business traveller/tourist'. In J.E.S. Higham and T.D. Hinch (eds) *Sport and Tourism: Globalisation, Mobility and Identity* (pp. 88–91). Oxford: Elsevier Butterworth Heinemann.

Hodges, J. and Hall, C.M. (1996) The housing and social impact of mega events: Lessons for the Sydney 2000 Olympics. Paper presented at the Proceedings 'Towards a More Sustainable Tourism', Dunedin, New Zealand, 3–6 December.

Hoffer, R. (1995) Down and out: On land, sea, air, facing questions about their sanity. *Sports Illustrated* 83 (1), 42–49.

Holden, A. (2000) Winter tourism and the environment in conflict: The case of Cairngorm, Scotland. *International Journal of Tourism Research* 2 (4), 247–260.

Hooper, I. (1998) The value of sport in urban regeneration: A case study of Glasgow. Paper presented at the Sport in the City Conference, Sheffield, UK, 2–4 July.

Hopkins, D. (2014) The sustainability of climate change adaptation strategies in New Zealand's ski industry: A range of stakeholder perceptions. *Journal of Sustainable Tourism* 22 (1), 107–126.

Hopkins, D. and Maclean, K. (2014) Climate change perceptions and responses in Scotland's ski industry. *Tourism Geographies: An International Journal of Tourism Space, Place and Environment* 16 (3), 400–414.

Hopkins, D. and Higham, J.E.S. (2016) *Low Carbon Mobility Transitions*. Oxford: Goodfellow Publishers.

Hopkins, D. and Higham, J.E.S. (2018) Climate change and tourism: Mitigation and global climate agreements. In C. Cooper, B. Gartner, N. Scott and S. Volo (eds) *The Sage Handbook of Tourism Management*. London: Sage.

Hopkins, D., Higham, J.E. and Becken, S. (2013) Climate change in a regional context: Relative vulnerability in the Australasian skier market. *Regional Environmental Change* 13 (2), 449–458.

Hopwood, B., Mellor, M. and O'Brien, G. (2005) Sustainable development: Mapping different approaches. *Sustainable Development* 13, 38–52.

Hornby, N. (1996) *Fever Pitch*. London: Cassel Group.

Horne, J. (1996) 'Sakka' in Japan. *Media, Culture & Society* 18 (4), 527–547.

Horne, W.R. (2000) Municipal economic development via hallmark tourism events. *The Journal of Tourism Studies* 11 (1), 30–35.

Host Broadcaster Consultancy (1997) *Critical Path Analysis for the 1998 Commonwealth Games*. Kuala Lumpur: Lambang Negara Malaysia.

Hritz, N. and Ross, C. (2010) The perceived impacts of sport tourism: An urban host community perspective. *Journal of Sport Management* 24 (2), 119–138.

Hsu, L-H. (2005) Revisiting the concept of sport. *Journal of Humanities and Social Sciences* 1 (2), 45–54.

Huber, N., Hergert, R., Price, B., Zäch, C., Hersoerger, A.M., Pütz, M., Kienast, F. and Bolliger, J. (2017) Renewable energy sources: Conflicts and opportunities in a changing landscape. *Regional Environmental Change* 17 (4), 1241–1255.

Hudson, S. (1999) *Snow Business: A Study of the International Ski Industry*. London: Cassell.

Hudson, S. (2002) The downhill skier in Banff National Park: An endangered species. In S. Gammon and J. Kurtzman (eds) *Sport Tourism: Principles and Practice* (pp. 89–110). Eastbourne: Leisure Studies Association.

Hudson, S. (2003) Winter sport tourism. In S. Hudson (ed.) *Sport and Adventure Tourism* (pp. 89–123). New York: The Haworth Hospitality Press.

Hudson, S. and Gilbert, D. (1998) Skiing constraints: Arresting the downhill slide. Paper presented at the Presentation at the Conference on Harnessing the High Latitudes, University of Surrey, Guildford, UK, 15–17 June.

Hudson, S. and Cross, P. (2005) Winter sports destinations: Dealing with seasonality. In J.E.S. Higham (ed.) *Sport Tourism Destinations: Issues, Opportunities and Analysis* (pp. 188–204). Oxford: Elsevier.

Hudson, S., Hinch T., Walker, G.J. and Simpson, B. (2010) Constraints to sport tourism: A cross-cultural analysis. *Journal of Sport & Tourism* 15 (1), 71–88.

Hudson, S. and Hudson, L. (2010) *Golf Tourism*. Oxford: Goodfellow Publishing.

Hudson, S. and Hudson, L. (2015) *Winter Sports Tourism*. Oxford: Goodfellow Publishers.

Hudson, S. and Hudson, L. (2016) The development and design of ski resorts: From theory to practice. In H. Richins and J. Hull (eds) *Mountain Tourism: Experiences, Communities, Environments and Sustainable Futures* (pp. 331–340). Wallingford: CABI.

Hultsman, W. (2012) Couple involvement in serious leisure: Examining participation in dog agility. *Leisure Studies* 31 (2), 231–253.

Humberstone, B. (2011) Embodiment and social and environmental action in nature-based sport: Spiritual spaces. *Leisure Studies* 30 (4), 495–512.

Humphreys, C. (2011) Who cares where I play? Linking reputation with the golfing capital and the implication for golf destinations. *Journal of Sport & Tourism* 16 (2), 105–128.

Humphreys, C. (2014) Understanding how sporting characteristics and behaviours influence destination selection: A grounded theory study of golf tourism. *Journal of Sport & Tourism* 19 (1), 29–54.

Humphreys, C.J. and Weed, M. (2014) Golf tourism and the trip decision-making process: The influence of lifestage, negotiation and compromise, and the existence of tiered decision-making units. *Leisure Studies* 33 (1), 75–95.

Hunter, C. (1995) Key concepts for tourism and the environment. In C. Hunter and H. Green (eds) *Tourism and the Environment: A Sustainable Relationship?* (pp. 52–92). London: Routledge.

Ifedi, F. (2008) *Sport Participation in Canada, 2005*. Ottawa: Statistics Canada.

Ingersoll, K.A. (2016) *Waves of Knowing: A Seascape Epistemology*. Durham, NC: Duke University Press.

Ingham, A.G., Howell, J.W. and Swetman, R.D. (1993) Evaluating sport 'hero/ines': Contents, forms, and social relations. *Quest* 45, 197–210.

Ingraham, C. (2016) Competition or exhibition? The Olympic arts and cultural policy rhetoric. *International Journal of Cultural Policy*, 1–16.

Inskeep, E. (1991) *Tourism Planning: An Integrated and Sustainable Development Approach*. New York: Van Nostrand Reinhold.

International Olympic Committee and World Tourism Organisation (2001) *Conclusions of the World Conference on Sport and Tourism*. Barcelona: International Olympic Committee and World Tourism Organization, Lausanne: International Olympic Committee.

IOC (2015) *Olympic Charter* (in force from 2 August 2016). Lausanne: International Olympic Committee.

Iordache, M.C. and Cebuc, I. (2009) Analysis of the impact of climate change on some European countries. *Analele Stiintifice ale Universitatii 'Alexandru Ioan Cuza' din Iasi* 56, 270–286. See http://anale.feaa.uaic.ro/anale/resurse/22_M03_Iordache_sa.pdf (accessed 24 April 2010).

Irwin, R. and Sandler, M. (1998) An analysis of travel behaviour and event induced expenditures among American collegiate championship patron groups. *Journal of Vacation Marketing* 4 (1), 78–90.

Iso-Ahola, S.E. (1982) Towards a social psychological theory for tourism motivation: A rejoinder. *Annals of Tourism Research* 12, 256–262.

Iso-Ahola, S.E. and Allen, J. (1982) The dynamics of leisure motivation: The effects of outcome on leisure needs. *Research Quarterly for Exercise and Sport* 53 (2), 141–149.

Jackson, E.L. (1989) Environmental attitudes, values and recreation. In E.L. Jackson and T.L. Burton (eds) *Understanding Leisure and Recreation: Mapping the Past, Charting the Future* (pp. 357–384). State College, PA: Venture Publishing.

Jackson, E.L. (1999) Leisure and the Internet. *Journal of Physical Education, Recreation & Dance* 70 (9), 18–22.

Jackson, E.L., Crawford, D.W. and Godbey, G. (1993) Negotiation of leisure constraints. *Leisure Sciences* 15 (1), 1–11.

Jackson, G. and Reeves, M. (1997) Evidencing the sport tourism relationship. In M.F. Collins and I.S. Cooper (eds) *Leisure Management: Issues and Applications* (pp. 172–188). Wallingford: CABI.

Jackson, S.J. (1994) Gretzky, crisis, and Canadian identity in 1988: Rearticulating the Americanization of culture debate. *Sociology of Sport Journal* 11 (4), 428–450.

Jackson, S.J. (1997) Sport, violence and advertising: A case study of global/local disjuncture in New Zealand. Paper presented at the North American Society for the Sociology of Sport Conference, Toronto, Canada, 5–8 November, 1997.

Jackson, S.J. (1998) The 49th paradox: The 1988 Calgary Winter Olympic Games and Canadian identity as contested terrain. In M. Duncan, G. Chich and A. Aycock (eds) *Player Culture Studies: Exploration in the Field of Play* (pp. 191–208). Greenwich: Ablex Publishing.

Jackson, S.J. and Andrews, D.L. (1999) Between and beyond the global and local: American popular sporting culture in New Zealand. In A. Yiannakis and M. Melnik (eds) *Sport Sociology: Contemporary Themes* (5th edn; pp. 467–474). Champaign, IL: Human Kinetics.

Jackson, S.J., Batty, R. and Scherer, J. (2001) Transnational sport marketing at the global/local nexus: The Adidasification of the New Zealand All Blacks. *International Journal of Sports Marketing and Sponsorship* 3 (2), 55–71.

Jackson, S.J. and McKenzie, A.D. (2000) Violence and sport in New Zealand. In C. Collins (ed.) Sport in New Zealand Society (pp. 153–170). Palmerston North, New Zealand: Dunmore Press.

Jamal, T. B. and Getz, D. (1994) Collaboration theory and community tourism planning. *Annals of Tourism Research* 22, 186–204.

Jang, S.S. (2004) Mitigating tourism seasonality: A quantitative approach. *Annals of Tourism Research* 31 (4), 819–836.

Japan National Tourism Organization (2017, 7 January) Press release. See http://www.jnto.go.jp/jpn/news/press_releases/pdf/170117_monthly.pdf (in Japanese).

Jefferson, A. (1986) Smoothing out the ups and downs in demand. *British Hotelier and Restaurateur* July/August, 24–25.

Jeffrey, D. and Barden, R.D. (2001) An analysis of the nature, causes and marketing implications of seasonality in the occupancy performance of English hotels. In T. Baum and S. Lundtorp (eds) *Seasonality and Tourism* (pp. 119–140). London: Pergamon.

Jennings, A. (1996) *The New Lords of the Rings: Olympic Corruption and How to Buy Gold Medals*. London: Pocket Books.

Jhally, S. (1989) Cultural studies and the sports/media complex. In L.A. Wenner (ed.) *Media, Sports and Society* (pp. 70–93). Newbury Park, CA: Sage.

Johnson, W.O. (1991) Sport in the year 2001: A fan's world. Watching sport in the 21st century. *Sports Illustrated* 75 (4), 40–48.

Johnston, C.S. (2001a) Shoring the foundations of the destination life cycle model, part 1: Ontological and epistemological considerations. *Tourism Geographies* 3 (1), 2–28.

Johnston, C.S. (2001b) Shoring the foundations of the destination life cycle model, part 2: A case study of Kona, Hawai'i Island. *Tourism Geographies* 3 (2), 135–164.

Jones, C. (2001) Mega-events and host-region impacts: Determining the true worth of the 1999 Rugby World Cup. *International Journal of Tourism Research* 3 (3), 241–251.

Jones, I. (2000) A model of serious leisure identification: The case of football fandom. *Leisure Studies* 19 (4), 283–298.

Jones, I. and Green, B.C. (2005) Serious leisure, social identity and sport tourism in sport. *Sport in Society* 8 (2), 164–181.

Kahanamoku, D. and Brennan, J. (1968) *Duke Kahanamoku's World of Surfing*. Sydney: Angus and Robertson.

Kane, M.J. and Zink, R. (2004) Package adventure tours: Markets in serious leisure careers. *Leisure Studies* 23 (4), 329–335.

Kang, Y.S. and Perdue, R. (1994) Long-term impacts of a mega-event on international tourism to the host country: A conceptual model and the case of the 1988 Seoul Olympics. *Journal of International Consumer Marketing* 6 (3/4), 205–226.

Kaplanidou, K. and Vogt, C. (2007) The interrelationship between sport event and destination image and sport tourists' behaviours. *Journal of Sport & Tourism* 12 (3–4), 183–206.

Kaspar, R. (1998) Sport, environment and culture. *Olympic Review* 20 (April/May), 1–5.

Keller, P. (2001) Sport and tourism: Introductory report. Paper presented at the World Conference on Sport and Tourism, Barcelona, Spain, 22–23 February.

Kennedy, E. and Deegan, J. (2001) Seasonality in Irish tourism, 1973–1995. In T. Baum and S. Lundtorp (eds) *Seasonality and Tourism* (pp. 119–140). London: Pergamon.

Kennelly, M., Lamont, M. and Moyle, B. (2015) Stories from the Sideline: Experiences of Serious Leisure Participants, Australian and New Zealand Academy of Leisure Sciences, 9–11 December, University of South Australia, Adelaide, Australia.

Kennelly, M., Moyle, B. and Lamont, M. (2013) Constraint negotiation in serious leisure: A study of amateur triathletes. *Journal of Leisure Research* 45(4), 466–484.

Kennelly, M. and Toohey, K. (2014) Strategic alliances in sport tourism: National sport organizations and sport tour operators. *Sport Management Review* 17 (4), 407–418.

Kennelly, J. and Watt, P. (2012) Seeing Olympic effects through the eyes of marginally housed youth: Changing places and the gentrification of East London. *Visual Studies* 27 (2), 151–160.

Kenyon, G. (1969) Sport involvement: A conceptual go and some consequences thereof. In G. Kenyon (ed.) *Aspects of Contemporary Sport Sociology* (pp. 77–100). Chicago, IL: Athletic Institute.

Kerr, A.K. and Emery, P.R. (2011) Foreign fandom and the Liverpool FC: A cyber-mediated romance. *Soccer and Society* 12 (6), 880–896.

Kerstetter, D. and Bricker, K. (2009) Exploring Fijian's sense of place after exposure to tourism development. *Journal of Sustainable Tourism* 17, 691–708.

Kiewa, J. (2002) Traditional climbing: Metaphor of resistance or metanarrative of oppression? *Leisure Studies* 21 (2), 145–161.

Kirkup, N. and Sutherland, M. (2017) Exploring the relationships between motivation, attachment and loyalty within sport event tourism. *Current Issues in Tourism* 20 (1), 7–14.

Klemm, M. and Rawel, J. (2001) Extending the school holiday season: The case of Europcamp. In T. Baum and S. Lundtorp (eds) *Seasonality in Tourism* (pp. 141–152). London: Pergamon.

Klenosky, D., Gengler, C. and Mulvey, M. (1993) Understanding the factors influencing ski destination choice: A means-end analytic approach. *Journal of Leisure Research* 25, 362–379.

Klostermann, C. and Nagel, S. (2014) Changes in German sport participation: Historical trends in individual sports. *International Review for the Sociology of Sport* 49 (5), 609–634.

Knott, B. (2015) The strategic contribution of sport mega-events to nation branding: The case of South Africa and the 2010 FIFA World Cup. Unpublished PhD thesis, Bournemouth University.

Koenig-Lewis, N. and Bischoff, E.E. (2005) Seasonality research: The state of the art. *International Journal of Tourism Research* 7 (4–5), 201–219.

Kotler, P., Haider, D.H. and Rein, I. (1993) Marketing Places: Attracting *Investment, Industry, and Tourism to Cities, States and Nations*. New York: The Free Press.

Krawczyk, Z. (1996) Sport as symbol. *International Review for the Sociology of Sport* 31 (4), 429–438.

Krein, K. (2008) Sport, nature and worldmaking. *Sports Ethics and Philosophy* 2 (3), 285–301.

Kreutzwiser, R. (1989) Supply. In G. Wall (ed.) *Outdoor Recreation in Canada* (pp. 19–42). Toronto: John Wiley & Sons.

Krippendorf, J. (1986) *The Holidaymakers: Understanding the Impact of Leisure and Travel*. London: Heinemann.

Krippendorf, J. (1995) Towards new tourism policies. In S. Medlik (ed.) *Managing Tourism* (pp. 307–317). Oxford: Butterworth Heinemann.

Kulczycki, C. and Hinch, T. (2014) 'It's a place to climb': Place meanings of indoor rock climbing facilities. *Leisure/Loisir* 38 (3–4), 271–293.

Kulczycki, C. and Hyatt, C. (2005) Expanding the conceptualization of nostalgia sport tourism: Lessons learned from fans left behind after sport franchise relocation. *Journal of Sport Tourism* 10 (4), 273–293.

Kurtzman, J. (1997) The peace games for the new millenium. *Journal of Sport Tourism* 4 (3), 24–27.

Kurtzman, J. (2001) Sport! tourism! culture! *Olympic Review* 27 (38), 20–27.

Kurtzman, J. and Zauhar, J. (1995) Tourism sport international council. *Annals of Tourism Research* 22 (3), 707–708.

Kurtzman, J. and Zauhar, J. (1997a) Sports tourism consumer motivation. *Journal of Sport Tourism* 4 (3), 17–30.

Kurtzman, J. and Zauhar, J. (1997b) A wave in time: The sports tourism phenomena. *Journal of Sport Tourism* 4 (2), 5–20.

Kurtzman, J. and Zauhar, J. (1998) Golf: A touristic venture. *Journal of Sport Tourism* 4 (4), 5–9.

Kyle, G. and Chick, G. (2007) The social construction of a sense of place. *Leisure Sciences* 29, 209–225.

Laidlaw, C. (1999) Sport and national identity: Race relations, business, professionalism. In B. Patterson (ed.) *Sport, Society and Culture in New Zealand* (pp. 11–18). Wellington, New Zealand: Victoria University Stout Research Centre.

Laidlaw, C. (2010) *Somebody Stole My Game*. New York: Hachette.

Lamont, M. (2014) Authentication in sports tourism. *Annals of Tourism Research* 45, 1–17.

Lamont, M., Kennelly, M. and Moyle, B. (2014) Costs and perseverance in serious leisure careers. *Leisure Sciences* 36 (2), 144–160.

Lamont, M., Kennelly, M. and Moyle, B. (2015) Non-participating entourage: The forgotten crowd in event management research? Working paper, Council for Australasian University Tourism and Hospitality Education (CAUTHE) Conference, Gold Coast, Australia, 2–5 February.

Lamont, M. and McKay, J. (2012) Intimations of postmodernity in sports tourism at the Tour de France. *Journal of Sport & Tourism* 17 (4), 313–331.

Laverie, D.A. and Arnett, D.B. (2000) Factors affecting fan attendance: The influence of identity salience and satisfaction. *Journal of Leisure Research* 32 (2), 225–246.

Law, A. (2001) Surfing the safety net: 'Dole bludging', 'surfies' and governmentality in Australia. *International Review for the Sociology of Sport* 36 (1), 25–40.

Law, C.M. (2002) *Urban Tourism: The Visitor Economy and the Growth of Large Cities.* London: Continuum.

Laws, E. (1991) *Tourism Marketing: Service and Quality Management Perspectives.* Cheltenham: Stanley Thornes Publishers.

Lawson, R., Tidwell, P., Rainbird, S., Loudon, N. and Della Bitta, P. (1996) *Consumer Behaviour in Australia and New Zealand.* Sydney: McGraw-Hill Book Company.

Lawson, R., Thyne, M. and Young, T. (1997) *New Zealand Holidays: A Travel Lifestyles Study.* Dunedin: The Marketing Department, University of Otago.

Lea, T., Young, M., Markham, F., Holmes, C. and Doran, B. (2012) Being moved (on) in Darwin and Alice Springs: Walking Australia's frontier towns. *Radical History Review* 114, 139–163.

Lealand, G. (1994) American popular culture and emerging nationalism in New Zealand. *The Phi Kappa Phi Journal* 74 (4), 34–37.

Lee, C., Bergin-Seers, S., Galloway, G., O'Mahony, B. and McMurray, A. (2008) Seasonality in the tourism industry: Impacts and strategies. CRC for Sustainable Tourism Pty Ltd.

Lee, J.J., Kyle, G. and Scott, D. (2012) Mediating effect of place attachment on the relationship between festival satisfaction and loyalty to the festival hosting destination. *Journal of Travel Research* 51 (6), 754–767.

Lefebvre, H. (1991) *The Production of Space* (D. Nicholson-Smith, trans.). Oxford: Blackwell. (Originally published 1974.)

Leiper, N. (1979) The framework of tourism: Towards a definition of tourism, tourist, and the tourist industry. *Annals of Tourism Research* 6 (4), 390–407.

Leiper, N. (1981) Towards a cohesive curriculum for tourism: The case for a distinct discipline. *Annals of Tourism Research* 8 (1), 69–74.

Leiper, N. (1990) Tourist attraction systems. *Annals of Tourism Research* 17 (3), 367–384.

Leisure Time (2002) Norway Cup. Publication 44. Bekkelagshogda 1109 Oslo, Norway.

Lenskyj, H.J. (1998) Sport and corporate environmentalism: The case of the Sydney 2000 Olympic Games. *International Review for the Sociology of Sport* 33 (4), 341–354.

Leonard, W.M. (1996) The odds of transiting from one level of sports participation to another. *Sociology of Sport Journal* 13 (3), 288–299.

Lesjø, J.H. (2000) Lillehammer 1994: Planning, figurations and the 'green' winter games. *International Review for the Sociology of Sport* 35 (3), 282–293.

L'Etang, J. (2006) Public relations and sport in promotional culture. *Public Relations Review* 32, 386–394.

Levey, B. (2010) It ain't fast food: An authentic climbing experience. In S.E. Schmid (ed.) *Climbing Philosophy for Everyone* (pp. 106–116). Chichester: Wiley-Blackwell.

Lew, A.A. (1987) A framework of tourist attraction research. *Annals of Tourism Research* 14 (3), 553–575.

Lew, A.A. (2001) Tourism and geography space. *Tourism Geographies* 3 (1), 1.

Lew, A.A. (2014) Introduction: Globalizing people, places, and markets in tourism. In A.A. Lew, C.M. Hall and A.M. Williams (eds) *The Wiley Blackwell Companion to Tourism* (pp. 191–196). Chichester: John Wiley & Sons.

Lewicka, M. (2011) Place attachment: How far have we come in the last 40 years? *Journal of Environmental Psychology* 31 (3), 207–230.

Lewis, G. and Redmond, G. (1974) *Sporting Heritage: A Guide to Halls of Fame, Special Collections and Museums in the United States and Canada.* South Brunswick, NJ: A.S. Barnes and Co., Inc.

Lima, G.N. and Morais, R. (2014) The influence of tourism seasonality on family business in peripheral regions (No. 03). Católica Porto Business School, Universidade Católica Portuguesa.

Liu, Z. (2003) Sustainable tourism development: A critique. *Journal of Sustainable Tourism* 11, 459–475.

Liverpool FC Supporters' Club Scandinavia (2002) Liverpool. See http://www.liverpool.no/ (accessed 25 September 2002).

Lockwood, A. and Guerrier, Y. (1990) Labour shortages in the international hotel industry. *Travel and Tourism Analyst* 6, 17–35.

Lopez Bonilla, J.M., Lopez Bonilla, L.M. and Sanz Altamira, B. (2006) Patterns of tourist seasonality in Spanish regions, *Tourism Planning & Development* 3 (3), 241–256.

Loverseed, H. (2000) Winter sports in North America. *Travel and Tourism Analyst* 6, 45–62.

Loverseed, H. (2001) Sports tourism in North America. *Travel and Tourism Analyst* 3, 25–41.

Lowes, M. and Awde, C. (2015) Sport tourism and the discourse of social cohesion at the world pond hockey championship event. *International Journal of Social Ecology and Sustainable Development (IJSESD)* 6 (2), 90–101.

Loy, J.W., McPherson, B.D. and Kenyon, G. (1978) *Sport and Social Systems: A Guide to the Analysis of Problems and Literature.* Reading: Addison Wesley.

Loy, J.W., McPherson, B.D. and Kenyon, G. (1978) Sport as a social phenomenon. In J.W. Loy, B.D. McPherson and G. Kenyon (eds) *Sport and Social Systems: A Guide to the Analysis of Problems and Literature* (pp. 3–26). Reading, MA: Addison-Wesley.

Lubowiecki-Vikuk, A.P. and Basirnska-Zych, A. (2011) Sport and tourism as elements of place branding: A case study on Poland. *Journal of Tourism Challenges & Trends* 4 (2), 33–52.

Lybrand, C.T.C.C. (1996) *Domestic Tourism Market Research Study.* Ottawa: Canadian Tourism Commission.

MacCannell, D. (1973) Staged authenticity: Arrangements of social space in tourist settings. *American Journal of Sociology* 79 (3), 589–603.

MacCannell, D. (1976) *The Tourists: New Theory of the Leisure Class.* New York: Schoken.

Magdalinski, T. (2000) The reinvention of Australia for the Sydney 2000 Olympic Games. *International Journal of the History of Sport* 17 (2/3), 304–322.

Maguire, J. (1994) Sport, identity politics, and globalization: Diminishing contrasts and increasing varieties. *Sociology of Sport Journal* 11 (4), 398–427.

Maguire, J. (1999) *Global Sport: Identities, Societies and Civilisations.* Cambridge: Polity Press.

Maguire, J. (2000) Sport and globalization. In J. Coakley and E. Dunning (eds) *Handbook of Sports Studies* (pp. 356–369). London: Sage.

Maguire, J. (2002) *Sport Worlds: A Sociological Perspective.* Champaign, IL: Human Kinetics.

Maguire, J. and Stead, D. (1996) Far pavilions? Cricket migrants, foreign sojourns and contested identities. *International Review for the Sociology of Sport* 31 (1), 1–24.

Maguire, J. and Stead, D. (1998) Border crossings: Soccer labour migration and the European Union. *International Review for the Sociology of Sport* 33 (1), 59–73.

Maier, J. and Weber, W. (1993) Sport tourism in local and regional planning. *Tourism Recreation Research* 18 (2), 33–43.

Maingard, J. (1997) Imag(in)ing the South African nation: Representations of identity in the Rugby World Cup 1995. *Theatre Journal* 49 (1), 15–28.

Manfredo, M.J. and Driver, B.L. (1983) A test of concepts inherent in experience based-settting management for outdoor recreation areas. *Journal of Leisure Research* 15 (3), 263–283.

Mannell, B., Walker, G.J. and Ito, E. (2014) Ideal affect, actual affect, and affect discrepancy during leisure and paid work. *Journal of Leisure Research* 46, 13–37.

Manzenreiter, W. (2008) The 'benefits' of hosting: Japanese experiences from the 2002 Football World Cup. *Asian Business and Management* 7, 201–224.

March, R. and Wilkinson, I. (2009) Conceptual tools for evaluating tourism partnerships. *Tourism Management* 30, 455–462.

Marciszewaski, B. (1998) Participation in free time sport recreation activities: Comparison of Gdansk Region, Poland and Guildford, United Kingdom. In S. Scraton (ed.) *Leisure Time and Space: Meanings and Values in People's Lives* (pp. 177–191). Eastbourne: Leisure Studies Association.

Marshall, N.A., Marshall, P.A., Abdulla, A., Rouphael, T. and Ali, A. (2011) Preparing for climate change: Recognising its early impacts through the perceptions of dive tourists and dive operators in the Egyptian Red Sea. *Current Issues in Tourism* 14 (6), 507–518.

Martín, J.M.M., Aguilera, J.D.D.J. and Moreno, V.M. (2014) Impacts of seasonality on environmental sustainability in the tourism sector based on destination type: An application to Spain's Andalusia region. *Tourism Economics* 20 (1), 123–142.

Mason, D.S. and Duquette, G.H. (2008) Urban regimes and sport in North American cities: Seeking status through franchises, events and facilities. *International Journal of Sport Management and Marketing* 3 (3), 221–241.

Mason, D.S., Duquette, G.H. and Scherer, J. (2005) Heritage, sport tourism and Canadian junior hockey: Nostalgia for social experience or sport place? *Journal of Sport Tourism* 10 (4), 253–271.

Mason, P. and Leberman, S. (2000) Local planning for recreation and tourism: A case study of mountain biking from New Zealand's Manawatu Region. *Journal of Sustainable Tourism* 8 (2), 97–115.

Mason, D., Ramshaw, G. and Hinch, T. (2008) Sports facilities and transnational corporations: Anchors of urban tourism development. In C.M. Hall and T. Coles (eds) *Tourism and International Business: Global Issues, Contemporary Interactions* (pp. 220–237). New York: Routledge.

Matheusik, M. (2001) When in doubt, shop. *Ski Area Management* 40 (1), 66–67, 83.

Mathieson, D. and Wall, G. (1987) *Tourism: Economic, Physical and Social Impacts.* London: Longman.

Matsumura, K. (1993) Sport and social change in the Japanese rural community. *International Review for the Sociology of Sport* 28 (2+3), 135–144.

May, V. (1995) Environmental implications of the 1992 Winter Olympic Games. *Tourism Management* 16 (4), 269–275.

McCabe, S. (2009) Who needs a holiday? Evaluating social tourism. *Annals of Tourism Research* 36 (4), 667–688.

McCabe, S. and Johnson, S. (2013) The happiness factor in tourism: Subjective well-being and social tourism. *Annals of Tourism Research* 41, 42–65.

McCabe, S., Joldersma, T. and Li, C. (2010) Understanding the benefits of social tourism: Linking participation to subjective well-being and quality of life. *International Journal of Tourism Research* 12 (6), 761–773.

McConnell, R. and Edwards, M. (2000) Sport and identity in New Zealand. In C. Collins (ed.) *Sport and Society in New Zealand* (pp. 115–129). Palmerston North, New Zealand: Dunmore Press.

McEnnif, J. (1992) Seasonality of tourism demand in the European community. *Travel and Tourism Analyst* 3, 67–88.

McGillivray, D. and Frew, M. (2015) From fan parks to live sites: Mega events and the territorialisation of urban space. *Urban Studies* 52 (14), 2649–2663.

McGuire, F.A. (1984) A factor analytic study of leisure constraints in advanced adulthood. *Leisure Sciences* 6, 313–326.

McGuirk, P.M. and Rowe, D. (2001) 'Defining moments' and refining myths in the making of place identity: The Newcastle Knights and the Australian Rugby League Grand Final. *Australian Geographical Studies* 39 (1), 52–66.

McIntosh, A.J. and Prentice, R.C. (1999) Affirming authenticity: Consuming cultural heritage. *Annals of Tourism Research* 26, 589–612.

McKay, J. and Kirk, D. (1992) Ronald McDonald meets Baron De Coubertin: Prime time sport and commodification. *Sport and the Media* Winter, 10–13.

Reaching Beyond the Gold: The Impact of the Olympic Games on Real Estate Markets. Chicago, IL: Jones Lang LaSalle IP, Inc.

McKenzie, D. (1998) Abreast in a boat: The race against breast cancer. *Canadian Medical Association Journal* 159 (4), 376–378.

McKercher, B. (1993) Some fundamental truths about tourism: Understanding tourism's social and environmental impacts. *Journal of Sustainable Tourism* 1 (1), 6–16.

McMurran, A. (1999, 22 January) More the 7000 expected for 2000 Games. *Otago Daily Times*, p. 18.

McPherson, B.D., Curtis, J.E. and Loy, J.W. (1989) *The Social Significance of Sport: An Introduction to the Sociology of Sport.* Champaign, IL: Human Kinetics Books.

Meinig, D. (1979) The beholding eye. In D. Meinig (ed.) *The Interpretation of Ordinary Landscapes* (pp. 33–48). New York: Oxford University Press.

Melbourne Sports and Aquatic Centre (2002) Facilities. See http://www.msac.com.au/sports.html (accessed 24 May 2002).

Melnick, M.J. and Jackson, S.J. (2002) Globalization American-style and reference idol selection: The importance of athlete celebrity others among New Zealand youth. *International Review for the Sociology of Sport* 37 (3–4), 429–448.

Melnick, M.J. and Loy, J.W. (1996) The effects of formal structure on leadership recruitment: An analysis of team captaincy among New Zealand provincial rugby teams. *International Review for the Sociology of Sport* 31 (1), 91–107.

Melnick, M.J. and Thomson, R.W. (1996) Segregation in New Zealand rugby football: A test of the anglocentric hypothesis. *International Journal of the History of Sport* 31 (2), 139–154.

Melo, R. and Sobry, C. (eds) (2017) *Sport Tourism: New Challenges in a Globalized World.* Newcastle upon Tyne: Cambridge Scholars Publishing.

Merkel, U., Lines, G. and McDonald, I. (1998) The production and consumption of sport cultures: Introduction. In U. Merkel, G. Lines and I. McDonald (eds) *The Production and Consumption of Sport Cultures: Leisure, Culture and Commerce* (Vol. Publication No. 62; pp. v–xvi). Eastbourne: Leisure Studies Association.

Metcalfe, A. (1993) The development of sporting facilities: A case study of East Northumberland, England, 1850–1914. *International Review for the Sociology of Sport* 28 (2+3), 107–119.

Meulen, V.D. and Salman, A.H.P.M. (1996) *Management of Mediterranean Coastal Dunes.* Amsterdam: Department of Physical Geography, University of Amsterdam.

Mihalik, B.J. and Simonetta, L. (1999) A midterm assessment of the host population's perceptions of the 1996 Summer Olympics: Support, attendance, benefits, and liabilities. *Journal of Travel Research* 37 (3), 244–248.

Millington, K., Locke, T. and Locke, A. (2001) Adventure travel. *Travel and Tourism Analyst* 4, 65–97.

Milne, S. and Ateljevic, I. (2004) Tourism economic development and the global–local nexus. In S. Williams (ed.) *Tourism: Critical Concepts in the Social Sciences* (pp. 81–103). London: Routledge.

Minnaert, L., Maitland, R. and Miller, G. (2009) Tourism and social policy: The value of social tourism. *Annals of Tourism Research* 36 (2), 316–334.

Miossec, J.M. (1977) L'image touristique comme introduction ý la gÈographie du tourisme. *Annales de gÈographie* 86, 473.

Miranda-Juan Andueza, J. (1997) The role of sport in the tourism destinations chosen by tourists visiting Spain. *Journal of Sport Tourism* 4 (3), 5–25.

Mitchell, L.S. and Murphy, P.E. (1991) Geography and tourism. *Annals of Tourism Research* 18 (1), 57–70.

Mitlin, D., Hickey, S. and Bebbington, A. (2007) Reclaiming development? NGOs and the challenge of alternatives. *World Development* 35 (10), 1699–1720.

Moen, J. and Fredman, P. (2007) Effects of climate change on Alpine skiing in Sweden. *Journal of Sustainable Tourism* 15 (4), 418–437.

Moore, M.S. (2011) *Sweetness and Blood: How Surfing Spread from Hawaii and California to the Rest of the World, with Some Unexpected Results*. New York: Rodale.

Moore, N.S.R. (1995) National mutual masters games, economic impact assessment, Dunedin, 5–13 February. Unpublished dissertation thesis, University of Otago.

Moragas Spa, M., Rivenburg, N.K. and Larson, J.F. (1995) *Television in the Olympics*. London: J. Libbey.

Morgan, M. (2007) 'We're not the barmy army!': Reflections on the sports tourist experience. *International Journal of Tourism Research* 9 (5), 361–372.

Morgan, N. (2014) Problematizing place promotion and commodification. In A.A. Lew, C.M. Hall and A.M. Williams (eds) *The Wiley Blackwell Companion to Tourism* (pp. 210–219). Chichester: John Wiley & Sons.

Morley, D. and Robins, K. (1995) *Spaces of Identity: Global Media, Electronic Landscapes and Cultural Boundaries*. London: Routledge.

Morse, J. (2001) The Sydney 2000 Olympic Games: How the Australian tourist commission leveraged the games for tourism. *Journal of Vacation Marketing* 7 (2), 101–107.

Moscardo, G. (2000) Cultural and heritage tourism: The great debates. In B. Faulkner, G. Moscardo and E. Laws (eds) *Tourism in the 21st Century: Lessons from Experience* (pp. 3–17). London: Continuum.

Moularde, J. and Weaver, A. (2016) Serious about leisure, serious about destinations: Mountain bikers and destination attractiveness. *Journal of Sport & Tourism* 20 (3,4), 285–304.

Mounet, J. and Chifflet, P. (1996) Commercial supply for river water sports. *International Review for the Sociology of Sport* 31 (3), 233–256.

Mourdoukoutas, P.G. (1998) Seasonal employment, seasonal unemployment and unemployment compensation. *American Journal of Economics and Sociology* 47 (3), 315–329.

Mowforth, M. (2002) *Tourism and Sustainability*. London: Routledge.

Mowforth, M. and Munt, I. (1998) *Tourism and Sustainability: New Tourism in the Third World*. London: Routledge.

Mowforth, M. and Munt, I. (2015) *Tourism and Sustainability: Development, Globalisation and New Tourism in the Third World*. London: Routledge.

Moyle, B.D., Kennelly, M. and Lamont, M. (2014) Risk management in triathlon: Amateur athletes reactions to the cancellation of an event. *International Journal of Event Management Research* 8 (1), 94–106.

Mules, T. (1998) Taxpayer subsidies for major sporting events. *Sport Management Review* 1 (1), 25–43.

Murata, S. (2010) Ekosupotsu ni yoru kankou kaihatsu no seitouka to sono riron: 'Seikatsu no umi' no jyuusouteki riyou wo meguru gyomin no taiou [Justification of tourism development by eco-sport, and its logic: Dealings of fishermen against overlapping activities on 'seikatsu no umi']. *Soshioroji* 55 (1), 21–38.

Murphy, P.E. (1985) *Tourism: A Community Approach*. New York: Methuen.

Murray, D. and Dixon, L. (2000) Investigating the growth of 'instant' sports: Practical implications for community leisure service providers. *The ACHPER Healthy Lifestyles Journal* 47 (3–4), 27–31.

Murray, J. (1996) How seasonality affects the economic viability of Canadian tourism businesses. In K. MacKay and K.R. Boyd (eds) *Tourism for All Seasons: Using Research to Meet the Challenge of Seasonality* (pp. 135–146). (Conference Proceedings of the Travel and Tourism Research Association.) Ottawa: Canada Chapter

Mykletun, R.J. and Vedø, K. (2002) BASE jumping in Lysefjord, Norway: A sustainable but controversial type of coastal tourism. Paper presented at the Tourism Research 2002, Cardiff, UK, 4–7 September.

Nadel, J.R., Font, A.R. and Roselló, A.S. (2004) The economic determinants of seasonal patterns. *Annals of Tourism Research* 31 (3), 679–711.

Nahrstedt, W. (2004) Wellness: A new perspective for leisure centres, health tourism, and Spas in Europe on the global health market. In K. Weiermair and C. Mathies (eds) *The Tourism and Leisure Industry: Shaping the Future* (pp. 181–198). Binghampton, NY: The Haworth Press.

Nash, R. and Johnston, S. (1998) The case of Euro96: Where did the party go? Paper presented at the Sport in the City Conference, Sheffield, UK, 2–4 July.

National Association of Sports Commissions (2017) Economic Impact. See www.sportscommissions.org/blog/economic-impact (accessed 6 January 2018).

Nauright, J. (1995) Introduction. In J. Nauright (ed.) *Sport, Power and Society in New Zealand: Historical and Contemporary Perspectives* (pp. 1–4). Sydney: University of New South Wales Printery.

Nauright, J. (1996) 'A besieged tribe'?: Nostalgia, white cultural identity and the role of rugby in a changing South Africa. *International Review for the Sociology of Sport* 31 (1), 69–89.

Nauright, J. (1997a) *Sport, Culture and Identities in South Africa*. London: Leicester University Press.

Nauright, J. (1997b) Masculinity, muscular Islam and popular culture: 'Coloured' rugby's cultural symbolism in working-class Cape Town c.1930–70. *International Journal of the History of Sport* 14 (1), 184–190.

Nettlefold, P.A. and Stratford, E. (1999) The production of climbing landscapes-as-texts. *Australian Geographical Studies*, 37 (2), 130–141.

Netto, A.P. (2009) What is tourism? Definitions, theoretical phases and principles. In J. Tribe (ed.) *Philosophical Issues in Tourism* (pp. 43–61). Bristol: Channel View Publications.

Nevo, I. (2000) Sport institutions and ideology in Israel. *Journal of Sport and Social Issues* 24 (4), 334–343.

New Zealand Rugby Almanac (1936–2017). Auckland: Upstart Press.

New Zealand Tourism Board (1998) All Blacks join forces with McCully, NZTB in South Africa, *Tourism News*, August, Wellington.

Nicholls, J. (1989) *The Competitive Ethos and Democratic Education*. Cambridge, MA: Harvard University Press.

Nogawa, H., Yamaguchi, Y. and Hagi, Y. (1996) An empirical research study on Japanese sport tourism in sport-for-all events: Case studies of a single-night event and a multiple-night event. *Journal of Travel Research* 35 (2), 46–54.

Nowak, J., Petit, S. and Sahli, M. (2009) Tourism and globalization: The international division of tourism production. *Journal of Travel Research* 47, 1–19.

Nusca, A. (2010) The future of air transport: Airbus unveils concept airplane for 2030. See http://www.zdnet.com/article/the-future-of-air-transport-airbus-unveils-concept-airplane-for-2030/ (accessed 19 July 2017).

O'Brien, D. (2006) Event business leveraging the Sydney 2000 Olympic games. *Annals of Tourism Research* 33 (1), 240–261.

O'Brien, D. (2007) Points of leverage: Maximizing host community benefit from a regional surfing festival. *European Sport Management Quarterly* 7 (2), 141–165.

O'Brien, D. and Chalip, L. (2007) Sport events and strategic leveraging: Pushing towards the triple bottom line. In A. Woodside and D. Martin (eds) *Tourism Management: Analysis, Behaviour, and Strategy* (pp. 318–338). Wallingford: CABI.

O'Reilly, N., Lyberger, M., McCarthy, L., Séguin, B. and Nadeau, J. (2008) Mega-special-event promotions and intent to purchase: A longitudinal analysis of the Super Bowl. *Journal of Sport Management* 22 (4), 392–409.

Oberti Resort Design (2016) Valemount Glacier Destination Master Plan. Valemount Glacier Destinations Ltd., Vancouver, BC.

Olds, K. (1998) Urban mega-events, evictions and housing rights: The Canadian case. *Current Issues in Tourism* 1 (1), 2–46.

Olympic Co-ordination Authority (1997a) *Environment: Committed to Conservation.* Homebush Bay, Sydney: Olympic Co-ordination Authority, New South Wales Government.

Olympic Co-Ordination Authority (1997b) *Environment: Protecting Nature's Gift.* Homebush Bay, Sydney: Olympic Co-Ordination Authority, New South Wales Government.

Olympic Co-Ordination Authority (1997c) *State of Play: A Report on Sydney 2000 Olympics Planning and Construction.* Homebush Bay, Sydney: Olympic Co-Ordination Authority, New South Wales Government.

Onkvisit, S. and Shaw, J.J. (1989) *Product Life Cycles and Product Management.* New York: Quorum Books.

Orams, M. (1999) *Marine Tourism: Development, Impacts and Management.* London: Routledge.

Orams, M. and Brons, A. (1999) Potential impacts of a major sport/tourism event: The America's Cup 2000. *Visions in Leisure and Business* 18 (1), 14–28.

Orsman, B. and Bingham, E. (2000, 27 October) America's Cup $640m boost to NZ economy. *New Zealand Herald*, p. 1.

Osborn, G. (2000) Football's legal legacy: Recreation, protest and disorder. In S. Greenfield and G. Osborn (eds) *Law and Sport in Contemporary Society* (pp. 51–68). London: F. Cass Publishers.

Osborne, A.C. and Coombs, D.S. (2013) Performative sport fandom: An approach to retheorizing sport fans. *Sport in Society* 16 (5), 672–681.

Otto, I. and Heath E.T. (2009) The potential contribution of the 2010 Soccer World Cup to climate change: An exploratory study among tourism industry stakeholders in the Tshwane Metropole of South Africa. *Journal of Sport Tourism* 14 (2/3), 169–191.

Oxford English Dictionary (2017) See https://en.oxforddictionaries.com/definition/sport (accessed on 28 August 2017).

Page, S.J., Brunt, P., Busby, G. and Connell, J. (2001) *Tourism: A Modern Synthesis.* London: Thomson Learning.

Page, S.J. and Hall, C.M. (2003) *Managing Urban Tourism.* Harlow: Pearson Education Ltd.

Panchal, J. (2014) *Tourism, Wellness and Feeling Good: Reviewing and Studying Asian Spa Experiences.* Abingdon: Routledge.

Paramio, J.L., Buraimo, B. and Campos, C. (2008) From modern to postmodern: The development of football stadia in Europe. *Sport in Society* 11 (5), 517–534.

Parrilla, J.C., Font, A.R. and Nadal, J.R. (2007) Accommodation determinants of seasonal patterns. *Annals of Tourism Research* 34 (2), 422–436.

Pavlovich, K. (2003) The evolution and transformation of a tourism destination network: The Waitomo Caves, New Zealand. *Tourism Management* 24 (2), 203–216.

Pawłowski, A. (2008) How many dimensions does sustainable development have? *Sustainable Development* 16, 81–90.

Pealo, W. and Redmond, G. (1999) Sport tourism: Moving into the new millennium. *Recreation and Parks BC* Spring, 22–24.

Pearce, D.G. (1987) *Tourism Today: A Geographical Analysis.* Harlow: Longman Scientific and Technical.

Pearce, D.G. (1989) *Tourism Development* (2nd edn). Harlow: Longman Scientific and Technical.

Pearce, D.G. and Butler, R.W. (eds) (1999) *Contemporary Issues in Tourism Development.* London: Routledge.

Pearce, P. (1982) *The Social Psychology of Tourist Behaviour.* Oxford: Pergamom Press.

Pearce, P. (1988) *The Ulysses Factor: Evaluating Visitors in Tourist Settings.* New York: Springer-Verlag.

Persson, C. (2002) The Olympic Games site decision. *Tourism Management* 23 (1), 27–36.

Pesqueux, Y. (2009) Sustainable development: A vague and ambiguous 'theory'. *Society and Business Review* 4, 231–245.

Pett, R. (2000) The end of the golden weather. Auckland Today, September, 7.

Pettersson, R. and Getz, D. (2009) Event experiences in time and space: A study of visitors to the 2007 World Alpine Ski Championships in Åre, Sweden. *Scandinavian Journal of Hospitality and Tourism* 9 (2–3), 308–326.

Picard, D. and Robinson, M. (eds) (2006) *Festivals, Tourism and Social Change: Remaking Worlds.* Clevedon: Channel View Publications.

Pickel-Chevalier, S., Violier, P. and Sari, N.P.S. (2016) Tourism and globalisation: Vectors of cultural homogenisation? (the case study of Bali). *Advances in Economics Business and Management Research* 19, 452–457.

Pickering, C., Castley, J. and Burtt, M. (2010) Skiing less often in a warmer world: Attitudes of tourists to climate change in an Australian ski resort. *Geographical Research* 48 (2), 137–147.

Pickmere, A. (2000, 27 October) A lot more than just a yacht race. *New Zealand Herald,* p. 13.

Pigeassou, C. (1997) Sport and tourism: The emergence of sport into the offer of tourism. Between passion and reason. An overview of the French situation and perspectives. *Journal of Sport Tourism* 4 (2), 20–38.

Pigeassou, C. (2002) Sport tourism as a growth sector: The French perspective. In S. Gammon and J. Kurtzman (eds) *Sport Tourism: Principles and Practice* (Vol. 76; pp. 129–140). Eastbourne: Leisure Studies Association.

Pigram, J.J. and Wahab, S. (1997) Sustainable tourism in a changing world. In S. Wahab and J.J. Pigram (eds) *Tourism, Development and Growth* (pp. 17–32). London: Routledge.

Pillay, U. and Bass, O. (2008) Mega-events as a response to poverty reduction: The 2010 FIFA World Cup and its urban development implications. *Urban Forum* 19 (3), 329.

Pitts, B.G. (1997) Sports tourism and niche markets: Identification and analysis of the growing lesbian and gay sports tourism industry. *Journal of Vacation Marketing* 5 (1), 31–50.

Plog, S. (1972) Why destination areas rise and fall in popularity. Paper presented at the Southern California Chapter of the Travel Research Bureau, San Diego, California, 10 October.

Poon, A. (1993) All-inclusive resorts. *Travel and Tourism Analyst* 2, 54–68.

Pope, S.W. (1997) Introduction: American sport history – toward a new paradigm. In S.W. Pope (ed.) *The New American Sport History: Recent Approaches and Perspectives* (pp. 1–30). Urbana, IL: University of Chicago.

Porteous, B. (2000) Sports development: Glasgow. *Leisure Manager* 18 (11), 18–21.

Porter, D. and Smith, A. (eds) (2013) *Sport and National Identity in the Post-War World.* London: Routledge.

Preuss, H. (2005) The economic impact of visitors at major multi-sport events. *European Sport Management Quarterly* 5 (3), 281–301.

Preuss, H. (2007) The conceptualisation and measurement of mega sport event legacies. *Journal of Sport & Tourism* 12 (3–4), 207–228.

Preuss, H. (2015) A framework for identifying the legacies of a mega sport event. *Leisure Studies* 34 (6), 643–664.

Priestley, G.K. (1995) Sports tourism: The case of golf. In G.J. Ashworth and A.G.J. Dietvorst (eds) *Tourism and Spatial Transformations: Implications for Policy and Planning* (pp. 205–223). Wallingford: CABI.

Priestley, G.K. (2006) Planning implications of golf tourism. *Tourism and Hospitality Research* 6 (3), 170–178.

Pujik, R. (2000) A global media event?: Coverage of the 1994 Lillehammer Olympic Games. *International Review for the Sociology of Sport* 35 (3), 309–330.

Pyo, S., Uysal, M. and Howell, R. (1988) Seoul Olympics visitor preferences. *Tourism Management* 9 (1), 68–72.

Pyo, S., Cook, R. and Howell, R.L. (1991) Summer Olympic tourist market. In S. Medlik (ed.) *Managing Tourism* (pp. 191–198). Oxford: Butterworth-Heinemann.

Ramshaw, G. (2010a) Remembering the rink: Hockey, figure skating and the development of community league recreation in Edmonton. *Prairie Forum* 35 (2), 27–42.

Ramshaw, G. (2010b) Living heritage and the sports museum: Athletes, legacy and the Olympic Hall of Fame and Museum, Canada Olympic Park. *Journal of Sport & Tourism* 15 (1), 45–70.

Ramshaw, G. (2011) The construction of sport heritage attractions. *Journal of Tourism Consumption and Practice* 3 (1), 1–25.

Ramshaw, G. (2014a) Sport, heritage, and tourism. *Journal of Heritage Tourism* 9 (3), 191–196.

Ramshaw, G. (2014b) Too much nostalgia? A decennial reflection on the heritage classic ice hockey event. *Event Management* 18 (4), 473–478.

Ramshaw, G. and Gammon, S. (2007) 'More than just nostalgia? Exploring the heritage/sport tourism nexus'. In S. Gammon and G. Ramshaw (eds) *Heritage, Sport and Tourism: Sporting Pasts – Tourist Futures* (pp. 9–22). London: Routledge.

Ramshaw, G. and Gammon, S. (2010) On home ground? Twickenham stadium tours and the construction of sport heritage. *Journal of Heritage Tourism* 5 (2), 87–102.

Ramshaw, G. and Gammon, S. (2015) Heritage and sport. In *The Palgrave Handbook of Contemporary Heritage Research* (pp. 248–260). Basingstoke: Palgrave Macmillan UK.

Ramshaw, G. and Gammon, S.J. (2016) Towards a critical sport heritage: Implications for sport tourism. *Journal of Sport & Tourism* 21 (2), 115–131.

Ramshaw, G. and Hinch, T. (2006) Place identity and sport tourism: The case of the heritage classic ice hockey event. *Current Issues in Tourism* 9 (4&5), 399–418.

Redmond, G. (1990) Points of increasing contact: Sport and tourism in the modern world. In A. Tomlinson (ed.) *Sport in Society: Policy, Politics and Culture* (pp. 158–167). Eastbourne: Leisure Studies Association.

Redmond, G. (1991) Changing styles of sports tourism: Industry/consumer interactions in Canada, the USA and Europe. In M.T. Sinclair and M.J. Stabler (eds) *The Tourism Industry: An International Analysis* (pp. 107–120). Wallingford: CABI.

Reeves, M.R. (2000) Evidencing the sport–tourism relationship: A case study approach. Unpublished PhD thesis, Loughborough University.

Reis, A.C., Sousa Mast, F.R. and Vieira, M.C. (2013) Public policies and sports in marginalised communities: The case of Cidade de Deus, Rio de Janeiro, Brazil. *World Leisure Journal* 55 (3), 229–251.

Reis, A.C., Sousa Mast, F.R. and Gurgel, L.A. (2014) Rio 2016 and the sport participation legacies. *Leisure Studies* 33 (5), 437–453.

Reis, A.C., Vieira, M.C. and Sousa Mast, F.R. (2016) 'Sport for development' in developing countries: The case of de Vilas Olímpicas do Rio de Janeiro, Brazil. *Sport Management Review* 19, 107–119.

Reisinger, Y. and Steiner, C.J. (2005) Reconceptualising object authenticity. *Annals of Tourism Research* 33, 65–86.

Relph, E. (1976) *Place and Placelessness*. London: Pion Limited.

Relph, E. (1985) Geographical experiences and being-in-the-world: The phenomenological origins of geography. In D. Seamon and R. Mugerauer (eds) *Dwelling, Place and Environment* (pp. 15–38). Dordrecht: Nijhoff.

Richards, G. (1996) Skilled consumption and UK ski holidays. *Tourism Management* 17, 25–34.

Ritchie, B.W., Shipway, R. and Cleeve, B. (2009) Resident perceptions of mega-sporting events: A non-host city perspective of the 2012 London Olympic Games. *Journal of Sport & Tourism* 14, 143–167.

Ritchie, J.B.R. (1984) Assessing the impact of hallmark events: Conceptual and research issues. *Journal of Travel Research* 13 (1), 2–11.

Ritchie, J.B.R. (1999) Policy formulation at the tourism/environment interface: Insights and recommendations from the Banff-Bow Valley study. *Journal of Travel Research* 38 (2), 100–110.

Ritchie, J.R.B. and Lyons, M. (1990) Olympulse VI: A post event assessment of resident reaction to te XV Olympic Winter Games. *Journal of Travel Research* Winter, 14–23.

Roberts, R. and Olson, J. (1989) *Winning is the Only Thing: Sports in America since 1945*. Baltimore, MD: The John Hopkins University Press.

Robins, K. (1991) Tradition and transition: National culture in its global context. In J. Corner and S. Harvey (eds) *Enterprise and Heritage* (pp. 21–44). London: Routledge.

Robins, K. (1997) What in the world is going on? In P. Du Gay (ed.) *Production of Culture/Cultures of Production* (pp. 11–67). London: Sage Publications.

Robinson, H. (1979) *A Geography of Tourism*. London: MacDonald and Evans.

Robinson, J.S. (2010) The place of the stadium: English football beyond the fans. *Sport in Society* 13 (6), 1012–1026.

Roche, M. (1994) Mega-events and urban policy. *Annals of Tourism Research* 21, 1–19.

Roche, M. (2000) *Mega-Events and Modernity: Olympics and Expos in the Growth of Global Culture*. London: Routledge.

Rodriguez-Diaz, J.A., Knox, J.W. and Weatherhead, E.K. (2007) Competing demands for irrigation water: Golf and agriculture in Spain. *Irrigation and Drainage* 56, 541–549.

Roehl, W., Ditton, R., Holland, S. and Perdue, R. (1993) Developing new tourism products: Sport fishing in the south-east United States. *Tourism Management* 14, 279–288.

Rogerson, C.M. (2014) Partnerships, tourism, and community impacts. In A.A. Lew, C.M. Hall and A.M. Williams (eds) *The Wiley Blackwell Companion to Tourism* (pp. 600–610). Chichester: John Wiley & Sons.

Rooney, J.F. (1988) Mega sports events as tourist attractions: A geographical analysis. Paper presented at the Tourism Research: Expanding the Boundaries. Travel and Tourism Research Association, Nineteenth Annual Conference, Montreal, Quebec.

Rooney, J.F. (1992) *Atlas of American Sport*. New York: Macmillan Publishing Co.

Rooney, J.F. and Pillsbury, R. (1992) Sports regions of America. *American Demographics* 14 (10), 1–10.

Ross, C.M. and Sharpless, D.R. (1999) Innovative information technology and its impact on recreation and sport programming. *Journal of Physical Education, Recreation & Dance* 70 (9), 26–30.

Ross, G.F. (1998) *The Psychology of Tourism*. Melbourne: Hospitality Press.

Ross, S.D. (2007) Segmenting sport fans using brand associations: A cluster analysis. *Sport Marketing Quarterly* 16 (1), 15.

Rowe, D. (1996) The global love-match: Sport and television. *Media, Culture & Society* 18, 565–582.

Rowe, D. and Lawrence, G. (1996) Beyond national sport: Sociology, history and postmodernity. *Sporting Traditions* 12 (2), 3–16.

Rowe, D., Lawrence, G., Miller, T. and McKay, J. (1994) Global sport? Core concern and peripheral vision. *Media, Culture and Society* 16, 661–675.

Rowe, D., McKay, J. and Miller, T. (1998) Come together: Sport, nationalism, and the media image. In L. Wenner (ed.) *Mediasport* (pp. 119–133). London: Routledge.

Royal and Ancient Golf Club of St. Andrews (R&A) (2010) Using water efficiently. See http://golfcoursemanagement.randa.org/en/Environmental-Impact/Using-water-efficiently.aspx (accessed 10 May 2017).

Ruskin, H. (1987) Selected views of socio-economic aspects of outdoor recreation, outdoor education and sport tourism. In M. Garmise (ed.) *Proceedings of the International Seminar and Workshop on Outdoor Education, Recreation and Sport Tourism* (pp. 18–37). Natanya: Emmanuel Gill Publishing.

Rutty, M., Matthews, L., Scott, D. and Del Matto, T. (2014) Using vehicle monitoring technology and eco-driver training to reduce fuel use and emissions in tourism: A ski resort case study. *Journal of Sustainable Tourism* 22 (5), 787–800.

Rutty, M., Scott, D., Steiger, R. and Johnson, P. (2014) Weather risk management at the Olympic Winter Games. *Current Issues in Tourism* 1–16. doi: 10.1080/13683500.2014.887665.

Rutty, M., Scott, D., Johnson, P., Pons, M., Steiger, R. and Vilella, M. (2017) Using ski industry response to climatic variability to assess climate change risk: An analogue study in Eastern Canada. *Tourism Management* 58, 196–204.

Ryan, C. (1995) *Researching Tourist Satisfaction: Issues, Concepts, Problems*. London: Routledge.

Ryan, C. and Lockyer, T. (2002) Masters' games—The nature of competitors' involvement and requirements. *Event Management* 7 (4), 259–270.

Ryan, C., Smee, A. and Murphy, S. (1996) Creating a database of events in New Zealand: Early results. *Festival Management and Event Tourism* 4 (3/4), 151–156.

Ryan, C. and Trauer, B. (2005) Sport tourist behaviour: The example of the Masters games. In J.E.S. Higham (ed.) *Sport Tourism Destinations: Issues, Opportunities and Analysis* (pp. 177–187). Oxford: Elsevier Butterworth Heinemann.

Saarinen, J. (2006) Traditions of sustainability in tourism studies. *Annals of Tourism Research* 33, 1121–1140.

Sage, G. H. (2016) *Globalizing Sport: How Organizations, Corporations, Media, and Politics are Changing Sport*. New York: Routledge.

Salazar, N. (2005) Tourism and glocalization: 'Local' tour guiding. *Annals of Tourism Research* 32 (3), 628–646.

Sallis, J.F., Cervero, R.B., Ascher, W., Henderson, K.A., Kraft, M.K. and Kerr, J. (2006) An ecological approach to creating active living communities. *Annual Review of Public Health* 27, 297–322.

Sampson, K.A. and Goodrich, C.G. (2009) Making place: Identity construction and community formation through 'sense of place' in Westland, New Zealand. *Society & Natural Resources* 22 (10), 901–915.

Sant, S.L. and Mason, D.S. (2015) Framing event legacy in a prospective host city: Managing Vancouver's Olympic bid. *Journal of Sport Management* 29 (1), 42–56.

Santana, G. (1998) Sports tourism and crisis management. *Journal of Sport Tourism* 4 (4), 9–22.

Saveriades, A. (2000) Establishing the social tourism carrying capacity for the tourist resorts of the east coast of the Republic of Cyprus. *Tourism Management* 21 (2), 147–156.

Scannell, L. and Gifford, R. (2010) Defining place attachment: A tripartite organizing framework. *Journal of Environmental Psychology* 30 (1), 1–10.

Schaffer, W. and Davidson, L. (1985) *Economic Impact of the Falcons on Atlanta: 1984.* Suwanee, GA: The Atlanta Falcons.

Schlossberg, H. (1996) *Sports Marketing.* Oxford: Blakewell.

Schollmann, A., Perkins, H.C. and Moore, K. (2001) Rhetoric, claims making and conflict in touristic place promotion: The case of central Christchurch, New Zealand. *Tourism Geographies* 3 (3), 300–325.

Schreyer, R. and Lime, D.W. (1984) A novice isn't necessarily a novice: The influence of experience use history on subjective perceptions of recreation participation. *Leisure Sciences* 6 (2), 131–149.

Schreyer, R., Lime, D.W. and Williams, D.R. (1984) Characterizing the influence of past experience on recreation behaviour. *Journal of Leisure Research* 16 (1), 34–50.

Schulenkorf, N. (2009) An ex ante framework for the strategic study of social utility of sport events. *Tourism and Hospitality Research* 9 (2), 120–131.

Schumacher, D.G. (2015) Report on the Sports Tourism Industry. National Association of Sports Commissions US.

Schuster, R.M., Thompson, J.G. and Hammitt, W.E. (2001) Rock climbers' attitudes toward management of climbing and the use of bolts. *Environmental Management* 28 (3), 403–412.

Scott, D. (2006a) Global environmental change and mountain tourism. In S. Gössling and C.M. Hall (eds) *Tourism and Global Environmental Change* (pp. 54–75). London/New York: Routledge.

Scott, D. (2006b) US ski industry adaptation to climate change: Hard, soft and policy strategies. In S. Gössling and C.M. Hall (eds) *Tourism and Global Environmental Change: Ecological, Social Economic and Political Interrelationships* (Chapter 15). Abingdon: Routledge.

Scott, D., Hall, C.M. and Gössling, S. (2016a) A review of the IPCC Fifth Assessment and implications for tourism sector climate resilience and decarbonization. *Journal of Sustainable Tourism* 24 (1), 8–30.

Scott, D., Hall, C.M. and Gössling, S. (2016b) A report on the Paris Climate Change Agreement and its implications for tourism: Why we will always have Paris. *Journal of Sustainable Tourism* 24 (7), 933–994.

Scott, D., Jones, B. and Konopek, J. (2007) Implications of climate and environmental change for nature-based tourism in the Canadian Rocky Mountains: A case study of Wateron Lakes National Park. *Tourism Management* 28, 570–579.

Scott, D., Jones, B., Lemieux, C., McBoyle, G., Mills, B., Svenson, S. and Wall, G. (2002) The Vulnerability of Winter Recreation to Climatic Change in Ontario's Lakelands Tourism Region. (Occasional Paper Number 18.) Waterloo, Ontario: Department of Geography Publication Series, University of Waterloo.

Scott, D. and McBoyle, G. (2007) Climate change adaptation in the ski industry. *Mitigation and Adaptation Strategies for Global Change* 12 (8), 1411.

Scott, D., McBoyle, G. and Minogue, A. (2007) Climate change and Quebec's ski industry. *Global Environmental Change* 17 (2), 181–190.

Scott, D., Steiger, R., Rutty, M. and Johnson, P. (2014) The future of the Olympic Winter Games in an era of climate change. *Current Issues in Tourism* 18 (10), 913–930. doi: 10.1080/13683500.2014.887664.

Selin, S. and Chavez, D. (1995) Developing an evolutionary tourism partnership model. *Annals of Tourism Research* 22 (4), 844–856.

Sell, B. (2000, 4 April) Sport in sport-mad New Zealand under severe strain. *New Zealand Herald*, p. A:3.

Sennet, R. (1999) Growth and failure: The new political economy and its culture. In M. Featherstone and S. Lash (eds) *Spaces of Culture: City–Nation–World* (pp. 14–26). London: Sage.

Shank, M.D. and Lyberger, M.R. (2014) *Sports Marketing: A Strategic Perspective*. New York: Routledge.

Shapcott, M. (1998) Commentary on 'Urban Mega-Events, Evictions and Housing Rights: The Canadian Case' by Chris Olds. *Current Issues in Tourism* 1 (2), 195–196.

Sharpley, R. (2014) Tourism: A vehicle for development. In R. Sharpley and D.J. Telfer (eds) *Tourism and Development: Concepts and Issues* (2nd edn, pp. 3–30). Bristol: Channel View Publications.

Sheard, K. (1999) A twitch in time saves nine: Birdwatching, sport, and civilising processes. *Sociology of Sport Journal* 16 (3), 181–205.

Sheard, R. (2014) *Sports Architecture*. Oxford: Taylor & Francis.

Sherlock, K. (2001) Revisiting the concept of hosts and guests. *Tourist Studies* 1 (3), 271–295.

Shibli, S. (1998) The economic impact of two major sporting events in two of the United Kingdom's 'National Cities of Sport'. Paper presented at the Sport in the City Conference, Sheffield, UK, 4–2 July.

Shimizu, S. (2014) Tokyo: Bidding for the Olympics and the discrepancies of nationalism. *International Journal of the History of Sport* 31 (6), 593–609.

Shipway, R. (2008) Road trip: Understanding the social world of the distance runner as sport tourist. In Proceedings of CAUTHE 2008 Annual Conference: Tourism and Hospitality Research, Training and Practice: 'Where the "bloody hell" are we?', Griffith University, Gold Coast, Australia, 11–14 February.

Shipway, R., Holloway, I. and Jones, I. (2012) Organisations, practices, actors and events: Exploring inside the distance running social world. *International Review for the Sociology of Sport* 48 (3), 259–276

Shipway, R. and Jones, I. (2007) Running away from home: Understanding visitor experiences and behaviour at sport tourism events. *International Journal of Tourism Research* 9 (5), 373–383.

Shipway, R., King, K., Lee, I.S. and Brown, G. (2016) Understanding cycle tourism experiences at the Tour Down Under. *Journal of Sport & Tourism* 20 (1), 21–39.

Shore, B. (1994) Marginal play: Sport at the borderlands of time and space. *International Review for the Sociology of Sport* 29 (4), 349–374.

Shultis, J. (2000) Gearheads and golems: Technology and wilderness recreation in the twentieth century. *International Journal of Wilderness* 6 (2), 17–18.

Silk, M. (2002) 'Bangsa Malaysia': Global sport, the city & the mediated refurbishment of local identities. *Media, Culture & Society* 25 (4), 775–794.

Silk, M. and Andrews, D.L. (2001) Beyond a boundary? Sport, transnational advertsing, and the reimaging of national culture. *Journal of Sport and Social Issues* 25 (2), 180–201.

Silk, M. and Jackson, S.J. (2000) Globalisation and sport in New Zealand. In C. Collins (ed.) *Sport in New Zealand Society* (pp. 99–113). Palmerston North, New Zealand: Dunmore Press.

Silvestre, G. and Oliveira, N.G. (2012) The revanchist logic of mega-events: Community displacement in Rio de Janeiro's West End. *Visual Studies* 27 (2), 204–210.

Simmons, D. and Urquhart, L. (1994) Measuring economic events: An example of endurance sports events. *Festival Management and Event Tourism* 2 (1), 25–32.

Simpson, J.A. and Weiner, E.S.C. (eds) (1989) *The Oxford English Dictionary* (2nd edn; Vol. XVII). Oxford: Clarendon Press.

Sims R., Schaeffer, R., Creutzig, F., Cruz-Núñez, X., D'Agosto, M., Dimitriu, D., Figueroa Meza, M.J., Fulton, L., Kobayashi, S., Lah, O., McKinnon, A., Newman, P., Ouyang, M., Schauer, J.J., Sperling, D. and Tiwari, G. (2014) Transport. In O. Edenhofer, R. Pichs-Madruga, Y. Sokona, E. Farahani, S. Kadner, K. Seyboth, A. Adler, I. Baum, S. Brunner, P. Eickemeier, B. Kriemann, J. Savolainen, S. Schlömer, C. von Stechow, T. Zwickel and J.C. Minx (eds) *Climate Change 2014: Mitigation of Climate Change. Contribution of Working Group III to the Fifth Assessment Report of the Intergovernmental Panel on Climate Change*. Cambridge, UK/New York: Cambridge University Press.

Slowikowski, S.S. and Loy, J.W. (1993) Ancient athletic motifs and the modern Olympic games: An analysis of rituals and representations. In A.G. Ingham and J.W. Loy (eds) *Sport in Social Development* (pp. 21–49). Champaign, IL: Human Kinetics.

Smith, A. (2000) Civil war in England: The clubs, the RFU, and the impact of professionalism on rugby union, 1995–99. In A. Smith and D. Porter (eds) *Amateurs and Professionals in Post-War British Sport* (pp. 146–188). London: Frank Cass Publishers.

Smith, A. (2005) Reimaging the city: The value of sport initiatives. *Annals of Tourism Research* 32 (1), 217–236.

Smith, A. (2010) The development of 'sports-city' zones and their potential value as tourism resources for urban areas. *European Planning Studies* 18 (3), 385–410.

Smith, S.L.J. (1983) *Recreation Geography*. London: Longman.

Snepenger, D., Houser, B. and Snepenger, M. (1990) Seasonality of demand. *Annals of Tourism Research* 17, 628–630.

Snyder, E. (1991) Sociology of nostalgia: Halls of fame and museums in America. *Sociology of Sport Journal* 8, 228–238.

Sonmez, S.F., Apolstolopoulos, Y. and Talow, P. (1999) Tourism in crisis: Managing the effects of terrorism. *Journal of Travel Research* 38, 13–18.

Sousa Mast, F.R., Reis, A.C., Gurgel, L.A. and Duarte, A.F.P.L.A. (2013) Are cariocas getting ready for the games? Sport participation and the Rio de Janeiro 2016 Olympic Games. *Managing Leisure* 18 (4), 331–335.

Sousa Mast, F.R., Reis, A.C., Sperandei, S., Gurgel, L., Vieira, M.C. and Pühse, U. (2016) Physical activity levels of economically disadvantaged women living in the Olympic city of Rio de Janeiro. *Women & Health*. doi: 10.1080/03630242.2015.1101745.

Spinney, J. (2006) A place of sense: A kinaesthetic ethnography of cyclists on Mont Ventoux. *Environment and Planning D: Society and Space* 24 (5), 707–732.

Spivack, S.E. (1998) Health spa development in the US: A burgeoning component of sport tourism. *Journal of Vacation Marketing* 4, 65–77.

Sport Tourism International Council (1995) Sport tourism categories revisited. *Journal of Sport Tourism* 2 (3), 1–4.

Sport Tourism International Council (1998) Case study of a sports tourism destination: Lake Placid and region. *Journal of Sport Tourism* 4 (4), 36–38.

Spracklen, K. (2013) *Leisure, Sports & Society*. Basingstoke: Palgrave Macmillan.

Spurr, R. (1999) Tourism. In R. Cashman and A. Hughes (eds) *Staging the Olympics: The Event and its Impact* (pp. 148–156). Sydney: UNSW Press.

Standeven, J. and De Knop, P. (1999) *Sport Tourism*. Champaign, IL: Human Kinetics.

Stanley, D. and Moore, S. (1997) Counting the leaves: The dimensions of seasonality in Canadian tourism. Paper presented at the Proceedings of the Travel and Tourism Research Association, Canadian Chapter, University of Manitoba, Winnipeg.

Stansfield, C.J. (1978) The development of modern seaside resorts. *Parks and Recreation* 5 (10), 14–46.

Stebbins, R.A. (2007) *Serious Leisure: A Perspective for Our Time* (Vol. 95). New Brunswick, NJ: Transaction Publishers.

Steele, W. (2006) Engaging rock climbers: Creating opportunities for collaborative planning and management in protected areas. *Australasian Parks and Leisure* 9 (2), 42–48.

Steenveld, L. and Strelitz, L. (1998) The 1995 Rugby World Cup and the politics of nation-building in South Africa. *Media, Culture & Society* 20 (4), 609–629.

Steiger, R. (2010) The impact of climate change on ski season length and snowmaking requirements n Tyrol, Austria. *Climate Research* 43 (3), 251–262.

Steiger, R. (2011a) The impact of climate change on ski touristic demand using an analogue approach. In K. Weiermair, H. Pechlahner, A. Strobl, M. Elmi and M. Schuckert (eds) *Coping with Global Climate Change. Strategies, Policies and Measures for the Tourism Industry* (pp. 247–256). Innsbruck: Innsbruck University Press.

Steiger, R. (2011b) The impact of snow scarcity on ski tourism. An analysis of the record warm season 2006/07 in Tyrol (Austria). *Tourism Review* 66 (3), 4–15.

Steiger, R. and Abegg, B. (2018) Ski areas' competitiveness in the light of climate change: Comparative analysis in the Eastern Alps. In D.K. Müller and M. Więckowski (eds) *Tourism in Transitions: Recovering Decline, Managing Change* (pp. 187–199). Cham: Springer.

Steiger, R. and Stötter, J. (2014) Climate change impact assessment of ski tourism in Tyrol. *Tourism Geographies* 15 (4), 577–600.

Stepchenkova, S. and Zhan, F. (2013) Visual destination images of Peru: Comparative content analysis of DMO and user-generated photography. *Tourism Management* 36, 590–601

Stevens, T. (1998) Capitalising on sport: Cardiff's future strategy. Paper presented at the Sport in the City Conference. Sheffield, UK, 2–4 July 1998.

Stevens, T. (2001) Stadia and tourism-related facilities. *Travel and Tourism Analyst* (2), 59–73.

Stevens, T. and van den Broek, M. (1997) Sport and tourism: Natural partners in strategies for tourism development. *Tourism Recreation Research* 22 (2), 1–3.

Stevens, T. and Wooton, G. (1997) Sports stadia and arena: Realising their full potential. *Tourism Recreation Research* 22 (2), 49–56.

Stevenson, D. (1997) Olympic Arts: Sydney 2000 and the Cultural Olympiad. *International Review for the Sociology of Sport* 32 (3), 227–238.

Stewart, B. (2001) Fab club. *Australian Leisure Management* October/November, 16–19.

Stewart, B. and Smith, A. (2000) Australian sport in a postmodern age. *International Journal of the History of Sport* 17 (2/3), 278–304.

Stewart, B., Smith, A. and Nicholson, M. (2003) Sport consumer typologies: A critical review. *Sport Marketing Quarterly* 12 (4), 206–216.

Stewart, J.J. (1987) The commodification of sport. *International Review for the Sociology of Sport* 22, 170–190.

Stranger, M. (2010) Surface and substructure: Beneath surfing's commodified surface. *Sport in Society* 13 (7–8), 1117–1134.

Su, C. (2014) From perpetual foreigner to national hero: A narrative analysis of US and Taiwanese news coverage of linsanity. *Asian Journal of Communication* 24 (5), 474–489.

Sugden, J. and Tomlinson, A. (1996) What's left when the circus leaves town? An evaluation of World Cup USA 1994. *Sociology of Sport Journal* 13 (3), 238–258.

SUKOM (1996, April) 'Let's Make it Great' (Issue No. 3, Restricted Circulation). Report to the General Assembly of the Commonwealth Games Federation. SUKOM 98 Berhad, Kuala Lumpur.

SUKOM (1998) Walkabout: Special edition for Kuala Lumpur 98. Kuala Lumpur: Malaysia-On-Call Sdn. Bhd./SUKOM 98 Berhad.

Swarbrooke, J. and Horner, S. (1999) *Consumer Behaviour in Tourism*. Oxford: Butterworth Heinemann.

Swart, K. (2000) An assessment of sport tourism curricular offerings at academic institutions. *Journal of Sport Tourism* 6 (1), 11–18.

Sylvester, C. (1999) The western idea of work and leisure: Traditions, transformations, and the future. In E.L. Jackson and T.L. Burton (eds) *Leisure Studies: Prospects for the Twenty-First Century* (pp. 17–33). State College, PA: Venture Publishing, Inc.

Tabata, R. (1992) Scuba diving holidays. In B. Weiler and C.M. Hall (eds) *Special Interest Tourism* (pp. 171–184). London: Belhaven Press.

Taks, M. (2013) Social sustainability of non-mega sport events in a global world. *European Journal for Sport and Society* 10 (2), 121–141.

Taks, M., Chalip, L. and Green, B.C. (2015) Impacts and strategic outcomes from non-mega sport events for local communities. *European Sport Management Quarterly* 15 (1), 1–6.

Taks, M., Chalip, L., Green, B.C., Kesenne, S. and Martyn, S. (2009) Factors affecting repeat visitation and flow-on tourism as sources of event strategy sustainability. *Journal of Sport & Tourism* 14 (2–3), 121–142.

Taks, M. and Scheerder, J. (2006) Youth sports participation styles and market segmentation profiles: Evidence and applications. *European Sport Management Quarterly* 6 (2), 85–121.

Tam, A. (1998) Critical success factors in sports tourism development: Their applicability to Singapore. *Journal of Sport Tourism* 5 (1), 16–26.

Tassiopoulos, D. and Haydam, N. (2008) Golf tourists in South Africa: A demand-side study of a niche market in sports tourism. *Tourism Management* 29 (5), 870–882.

Teigland, J. (1999) Mega-events and impacts on tourism: The predictions and realities of the Lillehammer Olympics. *Impact Assessment and Project Appraisal* 17 (4), 305–317.

Thamnopoulos, Y. and Gargalianos, D. (2002) Ticketing the large scale events: The case of Sydney 2000 Olympic Games. *Facilities* 20 (1/2), 22–33.

The Japan News/Yomiuri (2017, April 3) Japan to shoulder part of the security costs for Tokyo Olympic Games. See http://www.standard.net/World/2017/04/03/Japan-to-shoulder-part-of-security-costs-for-Tokyo-Olympic-Games (accessed 20 July 2017).

Thibault, L. (2009) Globalization of sport: An inconvenient truth. *Journal of Sport Management* 2, 1–20.

Thompson, S.M. (1985) Women in sport: Some participation patterns in New Zealand. *Leisure Studies* 4 (3), 321–331.

Thompson, S.M. (1988) Challenging the hegemony: New Zealand women's opposition to rugby and the reproduction of a capitalist patriarchy. *International Review for the Sociology of Sport* 23, 205–211.

Thompson, S.M. (1990) Thank the ladies for the plates. The incorporation of women into sport. *Leisure Studies* 9, 135–143.

Thomson, R. (2000) Physical activity through sport and leisure: Traditional versus non-competitive activities. *Journal of Physical Education New Zealand* 33 (1), 34–39.

Thornley, A. (2002) Urban regeneration and sports stadia. *European Planning Studies* 10 (7), 813–818.

Thorpe, H. (2011) *Snowboarding Bodies in Theory and Practice*. Basingstoke: Palgrave Macmillan.

Throssell, C., Lyman, G., Johnson, M., Stacey, G. and Brown, C. (2009) Golf course environmental profile measures water use, source, cost, quality, and management and conservation strategies. *Applied Turfgrass Science*. doi: 10.1094/ATS-2009-0129-01-RS.

Timothy, D. and Boyd, S.W. (2002) *Heritage Tourism*. London: Prentice Hall.

Todorova, V. (2015, February 28) Special report: Saving water in the UAE. The National UAE. See http://www.thenational.ae/uae/environment/special-report-saving-water-in-the-uae (accessed 24 June 2017).

Tokarski, W. (1993) Leisure, sports and tourism: The role of sports in and outside holiday clubs. In A.J. Veal, P. Jonson and G. Cushman (eds) *Leisure and Tourism. Social and Environmental Change* (pp. 684–686). University of Technology, Sydney: World Leisure and Recreation Association.

Tolkach, D., Chon, K.K. and Xiao, H. (2016) Asia Pacific tourism trends: Is the future ours to see? *Asia Pacific Journal of Tourism Research* 21 (10), 1071–1084.

Tomlinson, A. (1996) Olympic spectacle: Opening ceremonies and some paradoxes of globalization. *Media, Culture & Society* 18 (4), 583–602.

Tomlinson, A. (1999) *The Game's Up: Essays in the Cultural Analysis of Sport, Leisure and Popular Culture*. Aldershot: Ashgate Publishing Ltd.

Tomlinson, R., Bass, O. and Bassett, T. (2011) Before and after the vuvuzela: Identity, image and mega-events in South Africa, China and Brazil. *South African Geographical Journal* 93 (1), 38–48.

Tourist Authorities of Göteborg (2002) Gothia Cup. See http://www.gothiacup.se (accessed 24 October 2002).

Tow, S. (1994) Sports tourism: The benefits. *Journal of Sport Tourism* 2 (1), 1–7.

Traer, R. (2002) CEO: Canadian Sport Tourism Alliance. Personal communication, 1 November.

Trauer, B. and Ryan, C. (2005) Destination image, romance and place experience: An application of intimacy theory in tourism. *Tourism Management* 26 (4), 481–491.

Travel News (2010, 15 July) Sport tourism – Britain's great cash cow. See http://www.breakingtravelnews.com/news/article/sport-tourists-flock-to-britain/ (accessed 22 March 2017).

Tresidder, R. (1999) Tourism and sacred landscapes. In D. Crouch (ed.) *Leisure and Tourism Geographies* (pp. 137–152). London: Routledge.

Tsai, J.L. (2007) Ideal affect: Cultural causes and behavioral consequences. *Perspectives on Psychological Science* 2 (3), 242–259.

Tsai, J.L., Knutson, B. and Fung, H.H. (2006) Cultural variation in affect valuation. *Journal of Personality and Social Psychology* 90, 288–307.

Tuan, Y. (1974) *Topophilia: A Study of Environmental Perception, Attitudes, and Values*. Englewood Cliffs, NJ: Prentice Hall.

Tuan, Y.F. (1975) Place: An experiential perspective. *Geographical Review* 65 (2), 151–165.

Tuan, Y.F. (1977) *Space and Place: The Perspective of Experience*. Minneapolis, MN: University of Minnesota Press.

Tuck, J. (2003) Making sense of emerald commotion: Rugby union, national identity and Ireland. *Identities: Global Studies in Culture and Power* 10 (4), 495–515.

Tuppen, J. (2000) The restructuring of winter sports resorts in the French Alps: Problems, processes and policies. *International Journal of Tourism Research* 2 (5), 227–344.

Turco, D.M. (1999) Travelling and turnovers. Measuring the economic impacts of a street basketball tournament. *Journal of Sport Tourism* 5 (1), 6–11.

Turco, D.M., Riley, R. and Swart, K. (2002) *Sport Tourism*. Morgantown, WV: Fitness Information Technology.

UNFCCC (2015) Adoption of the Paris Agreement. See https://unfccc.int/resource/docs/2015/cop21/eng/l09r01.pdf (accessed 31 May 2016).

United Nations (2008) International Recommendations for Tourism Statistics 2008. Statistics Division Series M No. 83/Rev. 1, Department of Economic and Social Affairs, New York.

United Nations (2015) Transforming our World: The 2030 Agenda for Sustainable Development. See https://sustainabledevelopment.un.org/post2015/transformingourworld (accessed 10 September 2011).

United Nations (2017) Sports and Human Rights. See http://www.ohchr.org/EN/NewsEvents/Pages/SportsandHumanRights.aspx (accessed 9 September 2017).

United States Golf Association (USGA) (2014) Is your course environmentally and economically sound? See http://www.usga.org/course-care/usga-sustainability.html (accessed 23 February 2017).

Unruh, D. (1980) The nature of social worlds. *Pacific Sociological Review* 23, 271–296.

Upneja, A., Schafer, E.L., Seo, W. and Yoon, J. (2001) Economic benefit of sport fishing and angler wildlife watching in Pennsylvania. *Journal of Travel Research* 40 (1), 68–78.

Uriely, N. (1997) Theories of modern and postmodern tourism. *Annals of Tourism Research* 24 (4), 982–985.

Urry, J. (1990) *The Tourist Gaze: Leisure and Travel in Contemporary Societies*. London: Sage.

Usher, L.E. and Gomez, E. (2016) Surf localism in Costa Rica: Exploring territoriality among Costa Rican and foreign resident surfers. *Journal of Sport & Tourism* 20 (3,4), 195–216.

Van Wynsberghe, R. and Ritchie, I. (1998) *(Ir)Relevant Ring: The Symbolic Consumption of the Olympic Logo in Postmodern Media Culture*. Albany, NY: State University of New York Press.

Videira, N., Correia, A., Alves, I., Ramires, C., Subtil, R. and Martins, V. (2006) Environmental and economic tools to support sustainable golf tourism: The Algarve experience, Portugal. *Tourism and Hospitality Research* 6 (3), 204–217.

Vieira, M.C., Sperandei, S., Reis, A.C. and Silva, C.G.T. (2013) An analysis of the suitability of public spaces to physical activity practice in Rio de Janeiro. *Preventive Medicine* 57 (3), 198–200.

Vincent, J., Hill, J.S. and Lee, J.W. (2009) The multiple brand personalities of David Beckham: A case study of the Beckham brand. *Sport Marketing Quarterly* 18 (3), 173.

Voumard, S. (1995, 25 November 1995) Jonah's Big Date. *The Sydney Morning Herald*, p. 14.

Wahab, S. and Pigram, J.J. (eds) (1997) *Tourism Development and Growth: The Challenge of Sustainability*. London: Routledge.

Wahab, S. and Cooper, C. (2001) *Tourism in the Age of Globalisation*. London: Routledge.

Walker, G.J. and Virden, R.J. (2005) Constraints on outdoor recreation. In E. Jackson (ed.) *Constraints to Leisure* (pp. 201–219). State College, PA: Venture Publishing.

Walker, G.J., Deng, J. and Dieser, R.B. (2001) Ethnicity, acculturation, self-construal, and motivations for outdoor recreation. *Leisure Sciences* 23, 263–283.

Walker, G.J., Hinch, T. and Higham, J. (2010) Athletes as tourists: The roles of mode of experience and achievement orientation. *Journal of Sport & Tourism* 15 (4), 287–305.

Walker, S. (2001) Sport mad nation? *Australian Leisure Management* October/November, 32–35.

Wall, G. (1997) Sustainable tourism: Unsustainable development. In S. Wahab and J.J. Pigram (eds) *Tourism, Development and Growth* (pp. 33–49). London: Routledge.

Wall, G. and Mathieson, A. (2006) *Tourism: Change, Impacts, and Opportunities*. Harlow: Pearson Education.

Wang, N. (1999) Rethinking authenticity in tourism experience. *Annals of Tourism Research* 26 (2), 349–370.

Ward, T. (2009) Sport and national identity. *Soccer & Society* 10 (5), 518–531.

Warshaw, M. (2010) *The History of Surfing*. San Francisco, CA: Chronicle Books.

Washington, R.E. and Karen, D. (2001) Sport and society. *Annual Review of Sociology* 27, 187–212.

Watson, A.E. and Roggenbuck, J.W. (1991) The influence of past experience on wilderness choice. *Journal of Leisure Research* 23 (1), 21–36.

Weaver, D. and Lawton, L. (2002) *Tourism Management*. Brisbane: Wiley & Sons.

Webb, S. and Magnussen, B. (2002) Evaluating major sports events as cultural icons and economic drivers: A case study of Rugby World Cup 1999. Paper presented at Tourism Research 2002, an international interdisciplinary conference in Cardiff, Wales.

Wedemeyer, B. (1999) Sport and terrorism. In J. Riordan and A. Kruger (eds) *The International Politics of Sport in the 20th Century* (pp. 217–233). London: E&FN Spon.

Weed, M.E. (1999) 'More than sports holidays': An overview of the sport–tourism link. In M. Scarrott (ed.) *Exploring Sports Tourism: Proceedings of a SPRIG Seminar held at the University of Sheffield on 15 April 1999* (pp. 6–28). Sheffield: Sheffield Hallam University.

Weed, M.E. (2002) Football hooligans as undesirable sports tourists: Some meta-analytical speculations. In S. Gammon and J. Kurtzman (eds) *Sport Tourism: Principles and Practice* (pp. 35–52). Eastbourne: Leisure Studies Association.

Weed, M.E. (2003) Why the two won't tango! Explaining the lack of integrated policies for sport and tourism in the UK. *Journal of Sport Management* 17 (3), 258–283.

Weed, M.E. (2005) Research synthesis in sport management: Dealing with 'chaos in the brickyard'. *European Sport Management Quarterly* 5 (1), 77–90.

Weed, M.E. (2006) Sports tourism research 2000–2004: A systematic review of knowledge and a meta-evaluation of method. *Journal of Sport & Tourism* 11, 5–30.

Weed, M.E. (2007) *Olympic Tourism*. London: Routledge.

Weed, M.E. (2009) Progress in sports tourism research? A meta-review and exploration of futures. *Tourism Management* 30 (5), 615–628.

Weed, M.E. (2010) Sport fans and travel: Is 'being there' always important. *Journal of Sport & Tourism* 15, 103–109.

Weed, M.E. (2011) *The Journal of Sport & Tourism*: A maturing literature. In T.D. Hinch and J.E.S. Higham (eds) *Sport Tourism Development* (2nd edn; pp. 447–450). Bristol: Channel View Publications.

Weed, M.E. and Bull, C. (1997a) Integrating sport and tourism: A reviFpew of regional policies in England. *Progress in Tourism and Hospitality Research* 3 (2), 129–148.

Weed, M.E. and Bull, C. (1997b) Influences on sport tourism relations in Britain: The effects of government policy. *Tourism Recreation Research* 22 (2), 5–12.

Weed, M.E. and Bull, C. (1998) The search for a sport tourism policy network. In I. Cooper and M.F. Collins (eds) *Leisure Management: Issues and Applications* (pp. 277–298). Wallingford: CABI.

Weed, M.E. and Bull, C.J. (2004) *Sport Tourism: Participants, Policy and Providers*. Oxford: Butterworth Heinemann.

Weed, M.E. and Bull, C. (2009) *Sports Tourism: Participants, Policy and Providers* (2nd edn). Oxford: Butterworth-Heinemann.

Weed, M.E. and Bull, C. (2012) *Sports Tourism: Participants, Policy and Providers* (2nd edn). London: Routledge.

Weed, M.E., Bull, C., Brown, M., Dowse, S., Lovell, J., Mansfield, L. and Wellard, I. (2014) A systematic review and meta-analyses of the potential local economic impact of tourism and leisure cycling and the development of an evidence-based market segmentation. *Tourism Review International* 18 (1), 37–55.

Weighill, A.J. (2002) Canadian Domestic Sport Travel in 2001. Report prepared for Statistics Canada and the Canadian Tourism Commission, Ottawa, Canada.

Weiss, O., Norden, G., Hilscher, P. and Vanreusel, B. (1998) Ski tourism and environmental problems: Ecological awareness among different groups. *International Review for the Sociology of Sport* 33 (4), 367–380.

Wellard, I. (2016) *Researching Embodied Sport: Exploring Movement Cultures*. London: Routledge.

Wheaton, B. (2000) 'Just do it?': Consumption, commitment, and identity in the windsurfing subculture. *Sociology of Sport Journal* 17 (3), 254–274.

Wheaton, B. (ed.) (2004) *Understanding Lifestyle Sport: Consumption, Identity and Difference*. London: Routledge.

Wheaton, B. (2007) After sport culture: Rethinking sport and post-subcultural theory. *Journal of Sport and Social Issues* 31, 283–307.

Wheaton, B. (2013) *The Cultural Politics of Lifestyle Sports*. London: Routledge.

Wheeller, B. (1991) Tourism's troubled times: Responsible tourism is not the answer. *Tourism Management* June, 91–96.

Wheeler, K. and Nauright, J. (2006) A global perspective on the environmental impact of golf. *Sport in Society* 9 (3), 427–443.

Whistler Blackcomb (2017) See https://www.whistlerblackcomb.com/ (accessed 2 September 2017).

White, P. and Wilson, B. (1999) Distinctions in the stands. An investigation of Bourdieu's 'habitus', socioeconomic status and sport spectatorship in Canada. *International Review for the Sociology of Sport* 34 (3), 245–264.

Whitson, D. (2004) Bringing the world to Canada: 'The periphery of the centre'. *Third World Quarterly* 25 (7), 1215–1232.

Whitson, D. and Macintosh, D. (1996) The global circus: International sport, tourism and the marketing of cities. *Journal of Sport and Social Issues* 23, 278–295.

Wicker, P., Hallmann, K. and Breuer, C. (2013) Analyzing the impact of sport infrastructure on sport participation using geo-coded data: Evidence from multi-level models. *Sport Management Review* 16 (1), 54–67.

Wiley, C.E., Shaw, S.M. and Havitz, M.E. (2000) Men's and women's involvement in sports: An examination of the gendered aspects of leisure involvement. *Leisure Sciences* 22 (1), 19–31.

Williams, J. (1994) The local and the global in English soccer and the rise of satellite television. *Sociology of Sport Journal* 11 (4), 376–397.

Williams, D.R. and Champ, J.G. (2015) Performing leisure, making place: Wilderness identity and representation in online trip reports. In S. Gammon and S. Elkington (eds) *Landscapes of Leisure: Space, Place and Identities* (pp. 220–232). Basingstoke: Palgrave Macmillan.

Williams, D.R. and Kaltenborn, P. (1999) Leisure places and modernity: The use and meaning of recreational cottages in Norway and the USA. In D. Crouch (ed.) *Leisure/Tourism Geographies: Practices and Geographical Knowledge* (pp. 214–230). London: Routledge.

Williams, D.R., Patterson, M., Roggenbuck, J. and Watson, A. (1992) Beyond the commodity metaphor: Examining emotional and symbolic attachment to place. *Leisure Sciences* 14, 29–46.

Williams, A.M. and Shaw, G. (eds) (1988) *Tourism and Economic Development: Western European Experiences*. London: Belhaven.

Wilson, H. (1996) What is an Olympic city? Visions of Sydney 2000. *Media, Culture & Society* 18 (4), 603–618.

Wilson, H. (1998) Television's tour de force: The nation watches the Olympic Games. In D. Rowe and G. Lawrence (eds) *Tourism, Leisure and Sport: Critical Perspectives* (pp. 135–145). Sydney: Hodder Headline.

Wolbier, J. (2004) Matters of scale: Planet golf. *World Watch Magazine*, 17, np.

Wood, I. (1998, 2–4 July) Hong Kong: The event capital of Asia. Case studies on the International Dragon Boat Championships and Hong Kong Rugby Sevens. Paper presented at the Sport in the City, Sheffield, UK.

Woods, R. (2016) *Social Issues in Sport*. Champaign, IL: Human Kinetics.

Woodman, T. and Hardy, L. (2001) A case study of organizational stress in elite sport. *Journal of Applied Sport Psychology* 13 (2), 207–238.

Woolley-Fisher, P. and Chambers, E.J. (1990) The Edmonton Eskimos: An economic impact study (unpublished report). Western Centre for Economic Research, Edmonton, Canada.

World Commission on Environment and Development (WCED) (1987) *Our Common Future (The Bruntland Report)*. London: Oxford University Press.

World Tourism Organisation (1981) *Technical Handbook on the Collection and Presentation of Domestic and International Tourism Statistics*. Madrid: World Tourism Organization.

World Tourism Organisation (1994) *National and Regional Tourism Planning: Methodologies and Case Studies*. London: Routledge.

World Tourism Organisation (2001) *Tourism after 11 September 2001: Analysis, Remedial Actions and Prospects* (Special Report, Number 18, Market Intelligence Section). Madrid: World Tourism Organisation.

World Tourism Organisation (2002) *Tourism Recovery Already Underway*. Madrid: World Tourism Organisation.

World Tourism Organisation (2016) UNWTO tourism highlights, 2016 edition. See http://www.e-unwto.org/doi/pdf/10.18111/9789284418145 (accessed 25 January 2017).

World Tourism Organisation and International Olympic Committee (2001) Sport and Tourism: Sport Activities During the Outbound Holidays of the Germans, the Dutch and the French. Report published by the World Tourism Organisation and International Olympic Committee, Madrid.

Wright, J. and Clarke, G. (1999) Sport, the media and the construction of compulsory heterosexuality: A case study of women's rugby union. *International Review for the Sociology of Sport* 34 (3), 227–243.

Wynveen, C.J., Kyle, G.T. and Sutton, S.G. (2012) Natural area visitors' place meaning and place attachment ascribed to a marine setting. *Journal of Environmental Psychology* 32 (4), 287–296.

Xiang, Z. and Gretzel, U. (2010) Role of social media in online travel information search. *Tourism Management* 31 (2), 179–188.

Yamaguchi, S., Akiyoshi, R., Yamaguchi, Y. and Nogawa, H. (2015) Assessing the effects of service quality, past experience, and destination image on behavioral intentions in the spring training camp of a Japanese professional baseball team. *Journal of Convention & Event Tourism* 16 (3), 228–252.

Yang, L. and Wall, G. (2009) Ethnic tourism: A framework and an application. *Tourism Management* 30, 559–570.

Yeoman, I., Brass, D. and McMahon-Beattie, U. (2007) Current issue in tourism: The authentic tourist. *Tourism Management* 28, 1128–1138.

Yiannakis, A. (1975) A theory of sport stratification. *Sport Sociology Bulletin* 4, 22–32.

Yiannakis, A. and Gibson, H. (1992) Roles tourists play. *Annals of Tourism Research* 19 (2), 287–303.

Young, K. and Smith, M.D. (1988) Mass media treatment of violence in sport and its effects. *Current Psychology: Research & Reviews* 7 (4), 298–311.

Yusof, A. and Douvis, J. (2001) An examination of sport tourist profiles. *Journal of Sport Tourism* 6 (3), 1–10.

Zapata Campos, M.J. (2014) Partnerships, tourism, and community impacts. In A.A. Lew, C.M. Hall and A.M. Williams (eds) *The Wiley Blackwell Companion to Tourism* (pp. 567–577). Chichester: John Wiley & Sons.

Zhu, P. (2009) Studies on sustainable development of ecological sports tour resources and its industry. *Journal of Sustainable Development* 2, 80–83.

Index